Climate Change Solutions and Environmental Migration

This book lifts the taboo on maladaptation, a different driver of environmentally induced migration, which shines a light on the negative consequences arising from the solutions to climate change, adaptation and mitigation policies.

Through a systematic analysis and critique of existing mitigation and adaptation polices under the United Nations Framework Convention on Climate Change (UNFCCC) and international development community, and supplemented by a small empirical study in Indonesia, this book catalogues how maladaptation is manufactured under existing climate change solutions. It posits that customary communities in general – and women in particular – are disproportionately affected by the dominant market-driven logics that underscore current climate change solutions adopted by the UNFCCC. The injustice of maladaptation is highlighted as multi-faceted and explored using political, economic, social and ecological lenses, and the concept of environmental reintegration is also explored as a possible solution to this issue. Further possibilities are then presented in the *Afterword*, as a combination of what the new (post-neoliberalism) conjuncture could potentially look like.

This volume will be of great interest to students, scholars and practitioners of climate change, environmental policy, environmental migration and displacement, development studies, I/NGOs and civil society actors and activists more broadly.

Anna Ginty is an Industrial Adviser and Advocate for one of Australia's largest trade unions in Sydney, Australia, as well as a Visiting Fellow at the School of Social Sciences, Faculty of Arts & Social Sciences, University of New South Wales, Sydney.

Routledge Studies in Environmental Migration, Displacement and Resettlement

Resettlement Policy in Large Development Projects
Edited by Ryo Fujikura and Mikiyasu Nakayama

Global Implications of Development, Disasters and Climate Change
Edited by Susanna Price and Jane Singer

Repairing Domestic Climate Displacement
The Peninsula Principles
Edited by Scott Leckie and Chris Huggins

Climate Change Induced Migration and Human Rights
Law and Policy Perspectives
Edited by Andrew Baldwin, Dug Cubie, Dimitra Manou, Anja Mihr and Teresa Thorp

Migration and Environmental Change in the West African Sahel
Why Capabilities and Aspirations Matter
Victoria van der Land

Climate Refugees
Beyond the Legal Impasse?
Edited by Simon Behrman and Avidan Kent

Facilitating the Resettlement and Rights of Climate Refugees
An Argument for Developing Existing Principles and Practices
Avidan Kent and Simon Behrman

Communities Surviving Migration
Village Governance, Environment, and Cultural Survival in Indigenous Mexico
Edited by James P. Robson, Daniel Klooster and Jorge Hernández-Díaz

Climate Change Solutions and Environmental Migration
The Injustice of Maladaptation and the Gendered 'Silent Offset' Economy
Anna Ginty

For more information about this series, please visit: https://www.routledge.com/
Routledge-Studies-in-Environmental-Migration-Displacement-and-Resettlement/
book-series/RSEMDR.

Climate Change Solutions and Environmental Migration
The Injustice of Maladaptation and the Gendered 'Silent Offset' Economy

Anna Ginty

First published 2021
by Routledge
2 Park Square, Milton Park, Abingdon, Oxon OX14 4RN

and by Routledge
52 Vanderbilt Avenue, New York, NY 10017

Routledge is an imprint of the Taylor & Francis Group, an informa business

© 2021 Anna Ginty

The right of Anna Ginty to be identified as author of this work has been asserted by her in accordance with sections 77 and 78 of the Copyright, Designs and Patents Act 1988.

All rights reserved. No part of this book may be reprinted or reproduced or utilised in any form or by any electronic, mechanical, or other means, now known or hereafter invented, including photocopying and recording, or in any information storage or retrieval system, without permission in writing from the publishers.

Trademark notice: Product or corporate names may be trademarks or registered trademarks, and are used only for identification and explanation without intent to infringe.

British Library Cataloguing in Publication Data
A catalogue record for this book is available from the British Library

Library of Congress Cataloging-in-Publication Data
Names: Ginty, Anna, author.
Title: Climate change solutions and environmental migration : the injustice
    of maladaptation and the gendered 'silent offset' economy / Anna Ginty.
Description: Milton Park, Abingdon, Oxon ; New York, NY : Routledge,
    2021. | Series: Routledge studies in environmental migration, displacement
    and resettlement | Includes bibliographical references and index.
Identifiers: LCCN 2020047701 (print) | LCCN 2020047702 (ebook) | ISBN
    9780367490584 (hardback) | ISBN 9781003044277 (ebook)
Subjects: LCSH: Climate change mitigation. | Environmentalism.
Classification: LCC TD171.75 .G55 2021 (print) | LCC TD171.75 (ebook) |
    DDC 363.738/746–dc23
LC record available at https://lccn.loc.gov/2020047701
LC ebook record available at https://lccn.loc.gov/2020047702

ISBN: 978-0-367-49058-4 (hbk)
ISBN: 978-0-367-75522-5 (pbk)
ISBN: 978-1-003-04427-7 (ebk)

Typeset in Sabon
by Taylor & Francis Books

Contents

	List of figures	vi
	Acknowledgements	vii
1	Conceptualising key terms and their links: Maladaptation, adaptation, mitigation, environmental migration, gender and justice	1
2	Methodology: Critical, conceptual and empirical issues	16
3	Adaptation, development, maladaptation: Theory and practice	28
4	Mitigation and the Kyoto CDM: Manufacturing maladaptation	47
5	'Silent offsets' and feminist perspectives on women, climate change, UN-REDD+: Adapting to women	77
6	Findings of the Indonesian study	102
7	Where are the women?	120
8	Justice in the age of the Anthropocene: Reintegration as the fourth dimension of justice and the injustice of maladaptation	144
	Index	177

Figures

4.1 Simplified principle of the baseline (adapted from Michaelowa 2005). The emissions reduction gives the calculation for the amount of CO_2e (carbon dioxide equivalent) reduced as a result of the implementation of the project   57
4.2 Number of CDM projects in Asia by country as at 1 March 2011   61
4.3 Number of CDM projects in Asia by country as at 1 August 2020   61
4.4 Volume of CERs until 2012 in Asia by country as at 1 March 2011   62
4.5 Volume of CERs for Commitment Period 2 until 2020 in Asia by country as at 1 August 2020   62
4.6 Number of CDM projects in Latin America by country as at 1 March 2011   63
4.7 Number of CDM projects in Latin America by country as at 1 August 2020   63
4.8 Volume of CERs until 2012 in Latin America by country as at 1 March 2011   64
4.9 Volume of CERs for Commitment Period 2 until 2020 in Latin America by country as at 1 August 2020   64
5.1 Total productive system of an industrial society: Layer Cake with Icing   85
5.2 The Iceberg Model of capitalist patriarchal economics   86
5.3 Economic Dualism   87
8.1 Pivot to Extinction (author's own diagram)   171

Acknowledgements

Eileen Pittaway and the late Michael Wearing (PhD Supervisors), Linda Bartolomei, Claudia Tazreita, Jan Breckenridge, the Faculty of Arts & Social Science, University of New South Wales, Sydney for assistance with resources as a Visiting Fellow in writing this book. The three anonymous reviewers appointed by Routledge for their invaluable feedback on the book. Nik Kamvissis (who introduced me to 'flat ontology'), Stephanie Carmichael and Gabrielle Saulo (for digital drawings), Routledge editors Annabelle Harris and Matthew Shobbrook, from beginning to end.

Indonesia study: Anggalia Putri, HuMa in Jakarta, Bantaya in Palu, all anonymous Indonesian study participants in Jakarta, Palangkaraya and Palu who gave their time and kind hospitality. Rut Dini Prasti (my translator in Central Kalimantan) and her family, Tia, Harianson, Novi, Ilut and Deniss.

Family and friends: Monica Henley, Beth Cornwell, Amelia Weir, Tanya Ginty, T. Brett Ginty, Isabel Weir, Kim Barnfather, Isabella Angel, Lyn Pressman, Rob Jupp, Sarah Lendich, Christopher Kerle, John Grusovin, Gil Scrine, Simon Dodshon, Beth McKenzie, Steve Bonner.

Work colleagues: Julie Korlevska, Shimi Witkop, Kate Edmondson, Riz Chowdrey, Amelia Sereno, David Westrip, Ali Shaar, John McShane, Jacob K. Glud and Alex Kemp.

Mother Earth.

Thank you, one and all.

1 Conceptualising key terms and their links

Maladaptation, adaptation, mitigation, environmental migration, gender and justice

Introduction

Maladaptation in the context of climate change is a concept that has been little engaged with or understood outside the literature on geography, but is a key focus in this book, both in its links with climate change solutions and other key terms introduced in this overview. In 2010, geographers Jon Barnett and Saffron O'Neill in their seminal article, 'Maladaptation', defined it 'as an action taken ostensibly to avoid or reduce vulnerability to climate change that impacts adversely on, or increases the vulnerability of other systems, sectors or social groups' (Barnett & O'Neill 2010, p. 211). As 'one of the earliest attempts to systematically conceptualise maladaptation' (Juhola et al. 2016, p. 135), Barnett and O'Neill (2010) developed a typology of five pathways through which maladaptation can arise. In the absence of, or relative to, other alternatives, they suggest maladaptation arises when actions: '(1) increase emissions of greenhouse gases, (2) disproportionately burden the most vulnerable, (3) have high opportunity costs, (4) reduce incentives to adapt, and (5) set paths that limit the choices available to future generations [path dependency]' (Barnett & O'Neill 2010, p. 211).

Following this early attempt, others have joined in developing conceptual and practical resources to analytically strengthen and operationalise the concept, notably Magnan et al. (2016), in offering various frameworks to assess the risk of maladaptation through several dimensions. Their research of the current theoretical literature on maladaptation and practice-oriented insights from specific case studies 'shows that maladaptation is fundamentally a *process* that is influenced by *multiple drivers* and involves various *temporal and spatial scales*' (Magnan et al. 2016, p. 662). The chapter analyses in this book on both adaptation (Chapter 3) and mitigation (Chapter 4) policies confirm this; that is, not only can today's adaptation policy lead to tomorrow's maladaptation, so too can mitigation. Magnan et al. (2016) thus make a broad call for 'starting with the intention to avoid mistakes and not lock-in detrimental effects of adaptation-labelled initiatives ... thus advocates for the anticipation of the risk of maladaptation to become a priority for decision makers and stakeholders at large, from the international to the local scales' (Magnan et al. 2016, p. 646).

2 Conceptualising key terms and their links

However, it is Juhola et al.'s (2016) redefinition of maladaptation that captures this book's understanding of maladaptation, gained through desk-based conceptual research and empirical insights from an in-country study conducted in Indonesia in 2014–2015 investigating the impacts of the global United Nations Programme on Reducing Emissions from Deforestation and Forest Degradation (UN-REDD+) mitigation programme on customary communities, women specifically at the local project level. That is, 'maladaptation could be defined as a result of an intentional adaptation [mitigation] policy or measure directly increasing vulnerability for the targeted and/or external actors(s), and/or eroding preconditions for sustainable development by directly increasing society's vulnerability' (Juhola et al. 2016, p. 2649). While all three definitions 'acknowledge that maladaptation occurs when there are negative feedbacks that increase vulnerability' (Juhola et al. 2016), and all present typologies or frameworks to operationalise maladaptation emerging from adaptation actions, measures and/or policies, these frameworks for assessing maladaptation are exclusive to adaptation actions and not mitigation – the other side of the policy coin in climate change solutions that is explored herein.

This book then applies the maladaptation schemata, specifically Barnett and O'Neill's (2010) typology to mitigation policy, the Kyoto Protocol's Clean Development Mechanism (CDM) and, in so doing, agrees with Work et al. (2019) who 'take a critical approach to the concept of mitigation and invoke the term maladaptation to intervene into the persistent optimism of policy driven solutions' (Mosse 2005, cited in Work et al. 2019, p. 51). Additionally, these authors in following Taylor (2015, cited in Work et al. 2019), 'suggest that adaptation to climate change includes a broad array of institutional practices, discourses and policies and includes any and all activities created in the context of managing a changing climate (mitigation projects, for example)' (Work et al. 2019, p. 51). This modification opens up and positions mitigation under the critical lens of maladaptation which, following a detailed analysis of the CDM, presents a range of alternative policies obscured in the status quo of neoliberal market environmentalism shaping existing climate change solutions, or what some have called 'carbon capitalism' (Maxton-Lee 2018, 2020; Osborne 2018). Indeed, the 'gendered silent offset economy' discussed in Chapter 5 emerges from and is rooted in the asymmetrical power relations central to 'accumulation by decarbonisation' (Bumpus & Liverman 2008). That is, carbon capitalism's 'new' site of accumulation found in the solutions to climate change.

Key terms, concepts and their links to maladaptation: environmental migration, climate change solutions (adaptation and mitigation), gender and justice

The following pages in this conceptual overview introduce other key terms and concepts, in addition to maladaptation, as they are used, understood and interlinked in this book, including environmental migration, climate change solutions, gender and justice. Climate change solutions are messy, but what

Conceptualising key terms and their links 3

unites these concepts is embedded in the book's key, macro-level question that shapes and defines its normative thrust: What is the basis of just adaptation and mitigation policies that can be adopted by the United Nations Framework Convention on Climate Change (UNFCCC) once these have been subsumed within a global carbon market framework? Adopting a dual methodological framework (Chapter 2) using desk-based conceptual research, underscored by a critical analytic method and supplemented with feminist qualitative research in Indonesia in 2014–2015, maladaptation is used as a conceptual intervention to critique the solutions to climate change, adaptation and mitigation policies directed under the UNFCCC, where the analysis shows that market solutions to climate change framed and 'saturated in neoliberal ideas' (Hall 2011, p. 23) are manufacturing maladaptation, thus giving rise to the injustice of maladaptation, addressed in the final chapter of this book.

The analysis of adaptation and mitigation shows the neoliberal framing of climate change projects is not dissimilar to past agrarian policies and 'development-as-usual' projects (Maxton-Lee 2018, 2020; Vigil 2015, 2018; Work et al. 2019); findings congruent with those found in the Indonesian study. The upshot of Chapter 4's theoretical traverse in 'Mitigation and the Kyoto CDM' (and, which may serve as a cautionary warning for its successor, the new mitigation mechanism established under Article 6.4 of the Paris Agreement (UNFCCC Paris Agreement 2015)), leaves us with an international crediting mechanism meant to mitigate, but distinguishes itself in leading to maladaptation through all five pathways.

Cames et al.'s (2016) comprehensive analysis of the CDM 'suggests that the CDM still has fundamental flaws in terms of overall environmental integrity. It is likely that the large majority of the projects registered and CERs [Certified Emission Reductions] issued under the CDM are not providing real, measurable and additional emission reductions' (Cames et al. 2016, p. 11). Lacking in both environmental integrity and sustainable development, the dual objectives of the CDM, gives rise to a more serious question of, 'how then is adaptation to adapt to mitigation actions that have led to maladaptation?' The circularity of this question looks suspiciously similar to Sir Crispin Tickell (in Lovelock 2006, p. xiii) when pointing out Lovelock's summation of Gaia in a time of climate change: 'we are currently trapped in a vicious circle of positive feedback. What happens in one place very soon affects what happens in others.'

Environmental migration

An early focal point of this book, now discarded in its end point only, envisaged maladaptation arising from climate change solutions as a specific and dynamic driver of environmentally induced migration, potentially worthy of its own category, alongside those defining environmentally induced migration and displacement (Bronen 2009; Morinière 2009).[1] As two of the anonymous reviewers for this book's proposal correctly noted, 'climate migration has reached a certain maturity as a field with pluralism in approaches', while the

4 Conceptualising key terms and their links

other's main point of concern centred on 'the invention of (yet another) term for human mobility in an area of research rife with terminology: environmental migration/displacement, environmentally-induced migration/displacement, disaster displacement, climigration, trapped populations, and I am certain more terms that I have omitted here'. It was added that such a focal point could detract from the structural critique of climate change solutions leading to maladaptation this book makes.

In 2009, when research for this book began as a PhD thesis, a state-of-the-art literature overview on environmentally induced migration was published (Oliver-Smith & Shen 2009) outlining categories that, when tethered to Barnett and O'Neill's (2010) maladaptation typology published the following year, exposed a gap in the environmental change and migration literature; namely, maladaptation arising from market-based climate change solutions, adaptation and mitigation policies under the UNFCCC and Kyoto Protocol, presenting as a potentially different driver of migration/displacement.

Early research was emerging showing 'displacement linked to measures to mitigate or adapt to climate change. For example, biofuel projects and forest conservation could lead to displacement' (Norwegian Refugee Council Internal Displacement Monitoring Centre 2007, cited in Kolmannskog 2009). Less than a decade later, Sarah Vigil's (2015, 2018) research on land-grabbing and green-grabbing driven by two major climate change policies, biofuels and forest carbon projects, suggests 'there is a need to move beyond the category of environmentally induced migration displacement in order to include the impacts of climate change mitigation policies as a factor that influences displacement outcomes or migratory decisions' (Vigil 2015, p. 45). Later, she called for migration scholars 'to broaden the spectrum of their analysis' to include green-grabbing–induced displacement (Vigil 2018, p. 29).

Although Vigil (2015, 2018) does not use the term maladaptation, Work et al. (2019) do in their analysis of empirical research undertaken in Cambodia on projects associated with climate change mitigation and adaptation ('CCMA'): conservation, irrigation and reforestation. The authors state '[w]e did not explicitly seek data related to maladaptation on the ground … maladaptation became apparent as areas of concern through analysis' (Work et al. 2019, p. 50); indeed, so much so, its influence and priority is evident in the title of their important research, 'Maladaptation and development as usual?', joining other empirical studies highlighting maladaptation 'within this rapidly increasing research field' (Juhola et al. 2016, p. 135).

Maladaptation is a different driver to climate change itself, because it emerges as a negative consequence resulting from the solutions to climate change, therefore worthy – not as a distinct category in an already overcrowded environmentally induced migration schemata, which would only contribute more confusion than clarity, but as an analytic concept for the immediate attention of migration scholars and practitioners, as well as the international development community, who not only have to deal with human mobility as a result of climate change, but factor maladaptive climate change solutions into

Conceptualising key terms and their links 5

their migration analyses as well. In this way, the book contributes to conceptualising maladaptation as a first principle, and joins others in their support and advocacy, to place the risk of maladaptation as a priority on the planning agenda for decision makers (Magnan et al. 2016, p. 646). That is, not exclusively migration and international development scholars and practitioners, but for the Conference of the Parties (COP), 'the supreme decision-making body of the UNFCCC' (UNFCCC COP 2020) as well.

Climate change solutions – adaptation, mitigation and maladaptation

According to the leading scientific body for the assessment of climate change, the Intergovernmental Panel on Climate Change (IPCC), adaptation constitutes one of two human responses to climate change, the other being mitigation. That is, '[a] human intervention to reduce emissions or enhance the sinks of greenhouse gases' (IPCC 2018). Mitigation thus focuses on the source of climate change, while its consequences are addressed by adaptation (Schipper 2009). That is, adaptation refers to the 'adjustment in natural or human systems in response to actual or expected climatic stimuli or their effects, which moderates harm or exploits beneficial opportunities' (IPCC 2001, cited in Barnett & O'Neill 2010, p. 211).

By contrast, maladaptation, although not defined in the IPCC 2007 Fourth Assessment Report, was previously defined in the Third Assessment Report as 'an adaptation that does not succeed in reducing vulnerability but it increases it instead' (IPCC 2001, p. 990, cited in Barnett & O'Neill 2010, p. 211). By the Fifth Assessment Report, there are numerous references to the concept contained in the report, and a dedicated section entitled 'Addressing Maladaptation'. 'This suggests both that there is growing scholarship on maladaptation, and that the concept is making its way into the mainstream' (Magnan et al. 2016, p. 648). Juhola et al. (2016) agree, noting maladaptation's elevation is the result of the rapid pace of accumulating experiences of adaptation in recent years, where empirical studies have been highlighting maladaptation.

The analysis of adaptation in Chapter 3 focuses on the relatively 'new' salience of adaptation as a counterbalance to the decades-old 'bias toward mitigation', well served in the science and UNFCCC negotiating discussions literature on climate change (Schipper 2006; Pielke et al. 2007). 'In fact, during the first ten years the IPCC did not devote a single chapter of their reports to adaptation to climate change' (Hall 2016, p. 24). Critiquing adaptation does not imply a trade-off in mitigation policy because as Prins and Rayner (2007, p. 975) state, '[m]itigation and adaptation must go hand in hand'. Rather, foregrounding adaptation presages a shift to its matching term, maladaptation, where a more optimistic suggestion is made, that adaptation policies can be screened for potential maladaptation thus preventing, at least theoretically, its emergence. This approach embodies the precautionary principle which, like the principle of common but differentiated responsibilities, is one of two UNFCCC principles that are theoretically crucial but insufficiently implemented in

6 *Conceptualising key terms and their links*

practice (Elliott 2007). An appropriate place to critique adaptation thus begins with its various interpretations. 'Adaptation, development, maladaptation: theory and practice' builds upon how these interpretations have created 'conceptual confusion' around the term adaptation itself (Schipper 2009). It was noted in the early 1990s that 'the first obstacle to adaptation is reluctance to contemplate it' (Waggoner 1992, p. 146, cited in Schipper 2006, p. 86). On the analysis presented in this book, this could equally hold true for maladaptation.

According to Pielke et al. (2007), adaptation was understood in early policy discussions of climate change in the 1980s to be an important option for society. The shifting conceptual history of adaptation under the UNFCCC process (Schipper 2006) is explored further in Chapter 3 and introduced in brief here. According to Schipper (2006), adaptation in the 1970s and early 1980s was essentially an 'ecological' or 'individual' concept that put limits on the individual – for example, what are the ecological limits to human development and growth? Subsequent conferences, workshops and research over the following two decades 'contributed to a shift in interpretation of the concept, which has now [since the early 2000s] become associated with developing-country interests' (Schipper 2006, p. 87).

> Scientists engaged in drafting the UNFCCC who worked on adaptation issues, considered adaptive capacity to be a measure of the limits to responding to climate change, in line with the Club of Rome's 1972 report *The Limits to Growth* ... The focus was on how much a system could be stressed before it would collapse, an essentially ecological approach, although both ecosystems and human systems were considered.
>
> (Schipper 2006, p. 88)

Schipper (2006) states that the emphasis on 'tolerable limits', more precisely 'ecological limits' or 'limits of tolerance', would explain why adaptation became a secondary consideration to mitigation, which reflected an initial application of the term adaptive capacity in assessing how much *mitigation* was needed. Indeed, others have suggested that competing visions among the Conference of the Parties to the UNFCCC 'over who should participate in global mitigation efforts have stood in the way of establishing an effective climate change regime for nearly two decades' (Dimitrov 2010, Heywood 2007, cited in Mathur et al. 2014, p. 43). The bias toward mitigation over the following two decades, pushed adaptation policy to the margins as problematic for those advocating emissions reductions, that is mitigation. Chief among those opposing adaptation to climate change was the former US Vice-President Al Gore, who explained in 1992 his opposition to a belief in our ability to adapt to anything as representing the ultimate '"kind of laziness, an arrogant faith in our ability to react in time to save our skins"' (Gore 1992, cited in Pielke et al. 2007, p. 597).

As Pielke et al. (2007) go on to explain, perspectives have changed and the 'taboo on adaptation', reflected in the bias toward mitigation, can no longer be

Conceptualising key terms and their links   7

enforced due to several factors. Firstly, the timescale mismatch that exists. That is, irrespective of the decarbonisation of the global energy system, historical emissions dictate climate change is unavoidable. Whether we want it or not, adaptation to climate change will be required from us all (Adger et al. 2006). Secondly, for reasons other than greenhouse gas emissions, vulnerability to climate-related impacts is caused by a combination of increasing population, unsustainable development and socioeconomic inequity. Thirdly and finally, those who will be hardest hit and most vulnerable to climate change, developing countries, whose adaptive capacity is limited by financial constraints, are demanding greater attention to adaptation in international climate change policy negotiations (Pielke et al. 2007).

The IPCC's First Assessment Report, issued in 1990, served as a fundamental basis for climate change policy negotiations leading to the UNFCCC, which was adopted and opened for signature at the 1992 UN Conference for Environment and Development in Rio de Janeiro, informally known as the Earth Summit (UN Department of Public Information 1997; Röhr 2006). Article 2 of the Climate Convention states,

> [t]he ultimate objective of this Convention and any related legal instruments that the Conference of Parties may adopt is to achieve, in accordance with the relevant provisions of the Convention, stabilisation of greenhouse gas concentrations in the atmosphere at a level that would prevent dangerous anthropogenic interference with the climate system. Such a level should be achieved within a time frame sufficient to allow ecosystems to adapt naturally to climate change, to ensure that food production is not threatened and to enable economic development in a sustainable manner.
>
> (UNFCCC 1992)

Although adaptation is a term used in the Climate Convention five times, not once is it defined (Mace 2006; Ayers & Forsyth 2009). Thus, according to Mace (2006), adaptation must be understood in relation to terms that are defined in Article 1 under 'Definitions', such as 'adverse effects of climate change' and 'climate change' (UNFCCC 1992). In fact, Schipper (2006) argues that the reason why adaptation policy, as opposed to mitigation policy, developed so slowly stems, in part, from the lack of explicit provisions within the Climate Convention. Here, however, is not the place to detail how adaptation is discussed in the context of other articles that address related issues, 'particularly "developing-country issues" such as capacity building and technology transfer' (Schipper 2006, p. 90). That level of analysis is reserved throughout other chapters.

Suffice to say, the conceptual confusion about the meaning of adaptation is further problematised when incorporated into a development paradigm operating under a global carbon market. Marketised adaptation development (McMichael 2009), or what Ayers and Dodman (2010) typify as 'adaptation *plus* development', that is adaptation tacked onto development, and its

8 Conceptualising key terms and their links

attendant neoliberal watchwords, 'climate proofing development', 'main-streaming adaptation', 'climate resilient development' are explored. Adaptation may have begun its life as an ecological concept in the UNFCCC, but it has evolved to become synonymous with development (Schipper 2006). What emerges from the analysis between the adaptation and development policy nexus, is how potential maladaptation emerges from their conflation. In other words, viewing adaptation through a mainstream development paradigm is serving to exacerbate an already existing crisis in global development (Bakker & Silvey 2008).

Critiquing adaptation policy is to implicate mitigation, the 'bias towards mitigation' established under the Climate Convention (Schipper 2006), and lifting the 'taboo on adaptation' discourses (Pielke et al. 2007; Prins & Rayner 2007). The conceptual evolution of adaptation under the UNFCCC process explains how this bias was established, and how and why the focus turned to adaptation (Pielke 2009; Prins & Rayner 2007; Schipper 2006, 2009). Nordhaus made an early but correct observation in declaring 'mitigate we might; adapt we must' (Nordhaus 1994, p. 189, cited in Pielke 1998, p. 160).

Gender – women

Chapter 5's '"Silent offsets and feminist perspectives on women, climate change, UN-REDD+: adapting to women' presents a highly abstract, complex and theoretical account of women in the discourse on climate change, beginning with the early silence surrounding women in the climate change texts and negotiations, from the UNFCCC Secretariat to the Conference of the Parties (Carvajal-Escobar et al. 2008; Hemmati & Röhr 2009; Röhr 2006). The scientific framing of climate change as gender neutral by the IPCC (Röhr 2006) has profoundly shaped the way in which women's inclusion in adaptation and mitigation interventions has unfolded and this chapter presents that evolution.

Feminist analysis following the Indonesian study's women-specific findings, reveals the blunt instrument of gender mainstreaming has devolved from its once transformative vision of gender equality (Zalewski 2010) at the Beijing Women's Conference in 1995, to bland procedural participation via quotas in the UN-REDD+ project under study in Chapter 7's 'Where are the women?'

However, gender mainstreaming as the main instrument to achieve gender equality adopted at Beijing in 1995 had its roots two decades earlier. That is, it was not until 1975 that the UN started to take the 'woman question' seriously with the first international conference devoted to the 'second sex' held in Mexico that same year. Ever since, global gender equality has been an official UN priority (Bessis 2003). In her sweeping review of the UN and its agencies, Bessis (2003) maps the general patterns of international thinking about feminism, women and gender over the subsequent two and a half decades. In the review, she concludes that women's rights issues continue to be presented as 'the now familiar rhetoric on gender', thus risking further being an 'alibi for inaction' (Bessis 2003). This book uses both women and gender

Conceptualising key terms and their links 9

interchangeably, notwithstanding Bessis's (2003) valid criticism, but given the term's ubiquity in the literature.

What emerges from a critical analysis of the impacts of a REDD+ pilot project on customary communities (adat) in the Indonesian study, women in particular, is how they have been instrumentally used to meet the strategic interests of a global mitigation programme, far removed from their 'regime of familiarity' (Thevenot, cited by AP2 Jakarta 2014).[2] Conceptual and empirical insights thereby provide a meaning of what the notion of women as 'silent offsets' means – forest-dependent women and indigenous communities of the global South whose countries are participating partners in REDD+ are the prime subjects: embedded in a global economy that privileges the reproduction of carbon capital when, 'what matters is to move towards an economy that promotes a broader [social] reproduction of life' (Latin American Network of Women Transforming the Economy, cited in Filippini 2010). By chapter-end, what we are left with instead shows social reproduction is leveraged in the interests of carbon capitalism, giving rise to the broader gendered silent offset economy.

Introducing the book's key terms and concepts here demonstrates how, once defined, they are often better understood in their relationships with one another. That is, to speak of adaptation is to understand it relationally to mitigation, to maladaptation, and to both. A similar logic operates with non-foundational ways of theorising justice, such as Nancy Fraser's three-dimensional justice framework, introduced below, which is more interested in 'attaining *justice*, rather than attaining a sound *theory of justice*' (Schlosberg 2004, p. 520). Where claims for 'gender justice' and 'climate justice' are defined by reference to what 'starts not so much from a clear-sighted definition of what justice [is] but from widely shared intuitions of injustice' (Barnett 2010, p. 248).

Justice

What does the project for 'justice' mean once the atmospheric space has been gerrymandered by wealthy industrial countries at the expense of developing countries? The UNFCCC Kyoto CDM and UN-REDD+ Programme, as the analysis will show, are cases in point. That underlying question provoked the key question, addressed in the final chapter, 'Justice in the age of the Anthropocene: reintegration as the fourth dimension of justice and the injustice of maladaptation': 'What is the basis of just adaptation and mitigation policies that can be adopted by the UNFCCC once these have been subsumed within a global carbon market framework?'

The perversity of climate change solutions exacerbating vulnerability to the impacts of climate change – that is, maladaptation – in addition to climate change itself, adds another level of injustice for those most vulnerable to the impacts of climate change in the global South, those least responsible for historical greenhouse gas emissions who are set to bear a disproportionate burden of climate changes and are least able to pay. This double injustice, on its own,

10 *Conceptualising key terms and their links*

requires a deep critical analysis of the structural roots of mitigation and adaptation policies leading to the injustice of maladaptation.

Just as maladaptation's identification 'requires a broad systems perspective (Richards & Howden 2012), as well as a reflexive approach when identifying interdependencies and relationships between actors, sectors and goals' (Juhola et al. 2016, p. 138), so too does the intersection of maladaptation and justice. The conceptual intervention of Nancy Fraser's (2008) three-dimensional justice framework provides the scope necessary to assess justice claims-making associated with injustice emerging from the economic, cultural and political spheres of society. Overcoming injustice then requires dismantling the institutionalised obstacles to 'parity of participation', Fraser's (2005, 2008) most general meaning of social justice. That is, 'how fair or unfair are the terms of interaction that are institutionalised in the society? ... In my view, then, justice pertains *by definition* to social structures and institutional frameworks' (Fraser 2004, cited in Dahl et al. 2004, p. 378).

However, as Fraser's three-dimensional justice framework stands (with its corresponding injustices – economic redistribution (maldistribution), cultural recognition (misrecognition) and political representation (misrepresentation)), it is insufficient to accommodate the specific injustice of maladaptation whose remedy is found in environmental 'reintegration'. Hence, there is a need to reassess the scope of the tripartite framework to include this fourth dimension (reintegration/maladaptation), thus preserving the alliterative framework which characterises Fraserian justice.

Conclusion

This conceptual overview has introduced key terms and concepts as they are used, understood and linked in this book. The following chapter provides an overview of the methodology employed, which encompasses both desk-based conceptual research and feminist empirical qualitative research underscored by a critical analytic frame.

Notes

1 The categories that define environmentally induced migration have been proposed in a new convention regarding the naming of those fleeing environmental processes and events. The proposed new convention is the outcome of a joint effort between the United Nations University with the Environmental Change and Forced Migration Scenarios (EACH-FOR) (a co-financed research project within the Sixth Framework Programme of the European Commission project). The terms are: '"Environmental Emergency Migrants/Displacees" describes people who must flee because of rapid onset events and take refuge to save their lives; "Environmentally Forced Migrants" defines those who are compelled to leave to avoid gradual environmental deterioration and may not have a choice to return; "Environmentally Motivated Migrants" indicates those who choose to leave a deteriorating environment in order to avoid further weakening their livelihoods' (Renaud 2008, cited in Morinière 2009, p. 25).
2 For an explanation of the citation of participants in the Indonesian study, please see Chapter 6.

References

Adger, W, Paavola, J, Huq, S & Mace, M (eds) 2006, *Toward justice in adaptation to climate change: Fairness in adaptation to climate change*, MIT Press, Cambridge, Massachusetts, pp. 1–19.

Arora-Jonsson, S 2014, 'Forty years of gender research and environmental policy: where do we stand?', *Women's Studies International Forum*, vol. 47, pp. 295–308.

Ayers, J & Dodman, D 2010, 'Climate change adaptation and development I: the state of the debate', *Progress in Development Studies*, vol. 10, no. 2, pp. 161–168.

Ayers, J & Forsyth, T 2009, 'Community-based adaptation to climate change', *Environment: Science and Policy for Sustainable Development*, vol. 51, no. 4, pp. 22–31.

Bakker, I & Silvey, R 2008, 'Introduction: social reproduction and global transformations –from the everyday to the global' in I Bakker & R Silvey (eds), *Beyond states and markets: the challenges of social reproduction*, Routledge, London/New York, pp. 1–15.

Barnett, C 2010, 'Geography and ethics: justice inbound', *Progress in Human Geography*, vol. 35, no. 2, pp. 246–255.

Barnett, J & O'Neill, S 2010, 'Maladaptation', *Global Environmental Change*, vol. 20, pp. 211–213.

Bessis, S 2003, 'International organisations and gender: new paradigms and old habits', *Signs: Journal of Women in Culture and Society*, vol. 29, no. 2, pp. 633–647.

Boyd, E, Corbera, E & Estrada, M 2008, 'UNFCCC negotiations (pre-Kyoto to COP-9): what the process says about the politics of CDM-sinks', *International Environmental Agreements: Politics, Law and Economics*, vol. 8, no. 2, pp. 95–112.

Bronen, R 2009, 'Forced migration of Alaskan indigenous communities due to climate change: creating a human rights response', in A Oliver-Smith & X Shen (eds), *Linking environmental change, migration and social vulnerability*, UNU Institute for Environment and Human Security (UNU-EHS), e-book, accessed 21 May 2017, www.munichre-foundation.org/dms/MRS/Documents/Source2009_OliverSmith_Shen EnvironmentalChange_Migration.pdf.

Brooks, N, Grist, N & Brown, K 2009, 'Development futures in the context of climate change: challenging the present and learning from the past', *Development Policy Review*, vol. 27, no. 6, pp. 741–765.

Brown, K 2011, 'Sustainable adaptation: an oxymoron?', *Climate and Development*, vol. 3, no. 1, pp. 21–31.

Bumpus, A & Liverman, D 2008, 'Accumulation by decarbonisation and the governance of carbon offsets', *Economic Geography*, vol. 84, no. 2, pp. 127–155.

Burton, I 1997, 'Vulnerability and adaptive response in the context of climate and climate change', *Climatic Change*, vol. 36, no. 1, pp. 185–196.

Cames, M, Harthan, RO, Füssler, J, Lazarus, M, Lee, CM, Erickson, P & Spalding-Fecher, R 2016, *How additional is the Clean Development Mechanism?: analysis of the application of current tools and proposed alternatives*, Study prepared for DG CLIMA, Öko Institut, accessed 7 August 2020, https://ec.europa.eu/clima/sites/clima/ files/ets/docs/clean_dev_mechanism_en.pdf.

Carvajal-Escobar, Y, Quintero-Angel, M & Garcia-Vargas, M 2008, 'Women's role in adapting to climate change and variability', *Advances in Geosciences*, vol. 14, pp. 277–280.

Collett, L 2009, *A fair-weather friend? Australia's relationship with a climate-changed Pacific*, Australia Institute, accessed 21 May 2017, www.tai.org.au/node/1499.

12 *Conceptualising key terms and their links*

Conisbee, M & Simms, A 2003, *Environmental refugees: the case for recognition*, New Economics Foundation, London.

Dahl, H, Stoltz, P & Willig, R 2004, 'Recognition, redistribution and representation in capitalist global society: an interview with Nancy Fraser', *Acta Sociologica*, vol. 47, no. 4, pp. 374–382.

Dawson, A 2010, 'Climate justice: the emerging movement against green capitalism', *The South Atlantic Quarterly*, vol. 109, no. 2, pp. 313–338.

Dimitrov, R 2010a, 'Inside Copenhagen: the state of climate governance', *Global Environmental Politics*, vol. 10, no. 2, pp. 18–24.

Dimitrov, R 2010b, 'Inside UN climate change negotiations: the Copenhagen conference', *Review of Policy Research*, vol. 27, no. 6, pp. 795–821.

Elliott, L 2007, 'Improving the global environment: policies, principles and institutions', *Australian Journal of International Affairs*, vol. 61, no. 1, pp. 7–14.

Evans, S & Gabbatiss, J 2019a, 'In-depth Q & A: how "Article 6" carbon markets could "make or break" the Paris Agreement', *Carbon Brief*, accessed 8 August 2020, www.carbonbrief.org/in-depth-q-and-a-how-article-6-carbon-markets-could-make-or -break-the-paris-agreement.

Evans, S & Gabbatiss, J 2019b, 'COP25: Key Outcomes agreed at the UN climate talks in Madrid', *Carbon Brief*, accessed 8 August 2020, www.carbonbrief.org/cop25-key-outcomes-agreed-at-the-un-climate-talks-in-madrid.

Filippini, A 2010, 'Women and climate change in Cochabamba', *World Rainforest Movement*, no. 154, accessed 10 June 2017, http://wrm.org.uy/bulletins/issue-154/.

Fraser, N 2005, 'Reframing global justice in a globalising world', *New Left Review*, no. 36, Nov/Dec, pp. 69–88.

Fraser, N 2008, 'Reframing justice in a globalizing world', in N Fraser, *Scales of justice: reimagining political space in a globalizing world*, Polity Press, Cambridge, UK.

Global Environment Facility (GEF) Secretariat 2009, *Focal area: climate change*, Global Environment Facility, accessed 14 June 2017, www.thegef.org/sites/default/files/p ublications/ClimateChange-FS-June2009_2.pdf.

Hall, N 2016, *Displacement, development, and climate change: international organisations moving beyond their mandates*, Routledge.

Hall, S 2011, 'The neoliberal revolution: Thatcher, Blair, Cameron – the long march of neoliberalism continues', *Soundings: A journal of politics and culture*, no. 48, Summer, pp. 9–27.

Head, L 2010. 'Cultural ecology: adaptation-retrofitting a concept?', *Progress in Human Geography*, vol. 34, no. 2, pp. 234–242.

Hemmati, M & Röhr, U 2009, 'Engendering the climate-change negotiations: experiences, challenges, and steps forward', *Gender & Development*, vol. 17, no. 1, pp. 19–32.

Intergovernmental Panel on Climate Change (IPCC) 2007, *Climate Change 2007: Synthesis Report, Contribution of Working Groups I, II and III to the Fourth Assessment Report of the Intergovernmental Panel on Climate Change* (Core Writing Team, Pachauri, R.K and Reisinger, A. (eds.)), IPCC, Geneva, Switzerland, 104 pp.

Intergovernmental Panel on Climate Change (IPCC) 2014, *Climate Change 2014 Synthesis Report Summary for Policy Makers*, released 27 September 2013, accessed 15 June 2017, www.ipcc.ch/pdf/assessment-report/ar5/syr/AR5_SYR_FINAL_ SPM.pdf.

Intergovernmental Panel on Climate Change (IPCC) 2018, Annex I: Glossary [JBR Matthews, ed.]. In: *Global Warming of 1.5°C. An IPCC Special Report on the impacts of global warming of 1.5°C above pre-industrial levels and related global greenhouse gas emission pathways, in the context of strengthening the global response to the threat of climate change, sustainable development, and efforts to eradicate poverty* [V Masson-Delmotte, P Zhai, H-O Pörtner, D Roberts, J Skea, PR Shukla, A Pirani, W Moufouma-Okia, C Péan, R Pidcock, S Connors, JBR Matthews, Y Chen, X Zhou, MI Gomis, E Lonnoy, T Maycock, M Tignor and T Waterfield, eds.]. In Press, accessed 16 October 2020, www.ipcc.ch/sr15/chapter/glossary/.

Jones, R & Rahman, A 2007, 'Community-based adaptation', *Tiempo*, vol. 64, pp. 17–19.

Juhola, S, Glaas, E, Linner, B & Neset, T 2016, 'Redefining maladaptation', *Environmental Science & Policy*, vol. 55, January, pp. 135–140.

Kolmannskog, V 2009, *Climate change, disaster, displacement and migration: initial evidence from Africa*, UNHCR, The UN Refugee Agency Policy Development and Evaluation Service, accessed 15 June 2017, www.unhcr.org/4b18e3599.html.

Leopold, A & Mead, L 2009, 'Third international workshop on community-based adaptation to climate change', *Community Based Adaptation Climate Change Bulletin*, vol. 135, no. 2, accessed 10 June 2017, http://enb.iisd.org/download/pdf/sd/ymbvol135num2e.pdf.

Lovelock, J 2006, *The revenge of Gaia: why the earth is fighting back and how we can still save humanity*, Allen Lane, Camberwell.

Mace, M 2006, *Adaptation under the UN Framework Convention on Climate Change: the international legal framework*, MIT Press, Cambridge, Massachusetts.

Magnan, AK, Schipper, ELF, Burkett, M, Bharwani S, Burton, I, Erikson, S, Gemenne, F, Schaar, J & Zievogel, G 2016, 'Addressing the risk of maladaptation to climate change', *Wiley Interdisciplinary Reviews: Climate Change*, vol. 7, September/October, pp. 646–665.

Mathur, V, Afionis, S, Paavola, J, Dougill, A & Stringer, L 2014, 'Experiences of host communities with carbon market projects: towards multi-level climate justice', *Climate Policy*, vol. 14, no. 1, pp. 42–62.

Maxton-Lee, B 2018, 'Narratives of sustainability: a lesson from Indonesia: global institutions are seeking to shape an understanding of sustainability that undermines its challenge to their world view', *Sounding: A journal of politics and culture*, no. 70, Winter, pp. 45–57.

Maxton-Lee, B 2020, 'Forests, carbon markets, and capitalism: how deforestation in Indonesia became a geo-political hornet's nest', Guest Post, *REDD-Monitor*, accessed 22 August 2020, https://redd-monitor.org/2020/08/21/guest-post-forests-carbon-markets-and-capitalism-how-deforestation-in-indonesia-became-a-geo-political-hornets-nest/.

McAdam, J 2007, 'Climate change "refugees" and international law', *NSW Bar Association*, 24 October.

McAdam, J 2012, *Climate change, forced migration and international law*, Oxford University Press, Oxford.

McAfee, K 2014, *Green economy or buen vivir: can capitalism save itself*, Lunchtime Colloquium, Rachel Carson Centre for Environment and Society, online video, accessed 17 May 2020, www.carsoncenter.unimuenchen.de/events_conf_seminars/event_history/2014-events/2014_lc/index.html.

McGowan, A 2018, 'Life adrift: climate change, migration, critique', *Environment: Science and Policy for Sustainable Development*, vol. 60, no. 4, pp. 30–36.

McLeman, R & Gemenne, F 2018, 'Environmental migration research', in R McLeman & F Gemenne (eds), *Routledge Handbook of Environmental Displacement and Migration*, Routledge, London.

14   Conceptualising key terms and their links

McMichael, P 2009, 'Contemporary contradictions of the global development project: geopolitics, global ecology and the "development climate"', *Third World Quarterly*, vol. 30, no. 1, pp. 247–262.

Morinière, L 2009, 'Tracing the footprint of "environmental migrants" through 50 years of literature', in A Oliver-Smith & X Shen (eds), *Linking environmental change, migration and social vulnerability*, UNU Institute for Environment and Human Security (UNU-EHS), e-book, accessed 21 May 2017, www.munichre-foundation.org/dms/MRS/Documents/Source2009_OliverSmith_ShenEnvironmentalChange_Migration.pdf.

Nash, S 2019, *Negotiating migration in the context of climate change: international policy and discourse*, Bristol University Press, Bristol.

Okereke, C 2007, *Global justice and neoliberal environmental governance: ethics, sustainable development and international co-operation*, 1st edn, Taylor and Francis, Florence.

Oliver-Smith, A & Shen, X (eds) 2009, *Linking environmental change, migration and social vulnerability*, UNU Institute for Environment and Human Security (UNU-EHS), e-book, accessed 21 May 2017, www.munichre-foundation.org/dms/MRS/Documents.

Osborne, T 2018, 'The de-commodification of nature: indigenous territorial claims as a challenge to carbon capitalism', *Environment and Planning E: Nature and Space*, vol. 1, no. 1–2, pp. 25–75.

Pachauri, R 2009, *Statement of Dr. R. K. Pachauri Chairman, IPCC, UN Summit on Climate Change, 22 September 2009*, Intergovernmental Panel on Climate Change, accessed 10 June 2017.

Pielke, R 1998, 'Rethinking the role of adaptation in climate policy', *Global Environmental Change*, vol. 8, no. 2, pp. 159–170.

Pielke, R 2009, 'Rethinking the role of adaptation in climate policy', in ELF Schipper & I Burton (eds), *The Earthscan reader on climate change*, Earthscan, London, p. 345.

Pielke, R, Prins, G, Rayner, S & Sarewitz, D 2007, 'Climate change 2007: lifting the taboo on adaptation', *Nature*, vol. 445, no. 7128, pp. 597–598.

Prins, G & Rayner, S 2007, 'Time to ditch Kyoto', *Nature*, vol. 449, no. 7165, pp. 973–975.

Redman, J 2008, *World Bank: climate profiteer*, Sustainable Energy and Economy Network, Institute for Policy Studies, accessed 21 May 2017, www.ips-dc.org/wp-content/uploads/2008/04/1207774611WPCP-SEEN-April-9-08.pdf.

Robinson, M 2009, *Rule of law and climate justice*, The 2009 Griffith Lecture, 30 September 2009, accessed 10 May 2017, www.abc.net.au/radionational/programs/bigideas/the-2009-griffith-lecture-with-mary-robinson/3099570#transcript.

Röhr, U 2006, 'Gender relations in international climate change negotiations', *Berlin: LIFE eV/genanet*.

Rooke, A 2009, 'Doubling the damage: World Bank climate investment funds undermine climate and gender justice', *Gender Action*, accessed 10 May 2017, www.genderaction.org/images/2009.02_Doubling%20Damage_AR.pdf.

Scheraga, J & Grambsch, A 1998, 'Risks, opportunities, and adaptation to climate change', *Climate Research*, vol. 11, no. 1, pp. 85–95.

Schipper, ELF 2006, 'Conceptual history of adaptation in the UNFCCC process', *Review of European Community & International Environmental Law*, vol. 15, no. 1, pp. 82–92.

Schipper, ELF 2009, 'Conceptual history of adaptation in the UNFCCC process', in ELF Schipper & I Burton (eds), *The Earthscan reader on climate change*, Earthscan, London, pp. 359–376.

Schlosberg, D 2004, 'Reconceiving environmental justice: global movements and political theories', *Environmental Politics*, vol. 13, no. 3, pp. 517–540.

Stern, N 2006, *Stern review: the economics of climate change, summary of conclusions*, HM Treasury, UK Government Web Archive, accessed 10 May 2017, http://webarchive.nationalarchives.gov.uk/20130129110402/http://www.hm-treasury.gov.uk/d/CLOSED_SHORT_executive_summary.pdf.

United Nations Department of Public Information 1997, *UN Conference on Environment and Development (1992)*, United Nations, accessed 10 May 2017, www.un.org/geninfo/bp/enviro.html.

United Nations Framework Convention on Climate Change (UNFCCC) 1992, *United Nations General Assembly*, 20 January 1994, accessed 9 May 2017, www.refworld.org/docid/3b00f2770.html.

United Nations Framework Convention on Climate Change (UNFCCC) 2015, *Paris Agreement*, accessed 16 September 2020, https://unfccc.int/process-and-meetings/the-paris-agreement/the-paris-agreement.

United Nations Framework Convention on Climate Change (UNFCCC) Conference of Parties (COP) 16 AWG-LCA 2010, *Report on the Conference of Parties on its sixteenth session, held in Cancun from 29 November to 10 December*, United Nations Framework Convention on Climate Change, accessed 15 June 2017, http://unfccc.int/resource/docs/2010/cop16/eng/07a01.pdf#page=2.

United Nations Framework Convention on Climate Change (UNFCCC) Conference of Parties (COP) 2020, *Bodies: Conference of the Parties (COP)*, United Nations Climate Change, accessed 3 October 2020, https://unfccc.int/process/bodies/supreme-bodies/conference-of-the-parties-cop#:~:text=UNFCCC%20Nav,Conference%20of%20the%20Parties%20(COP).

Vigil, S 2015, 'Displacement as a consequence of climate change mitigation policies', *Forced Migration Review*, no. 49, May, pp. 43–45.

Vigil, S 2018, 'Green grabbing-induced displacement', in R McLeman & F Gemenne (eds), *Routledge Handbook of Environmental Displacement and Migration*, Routledge, London.

Work, C, Rong, V, Song, D & Scheidel, A 2019, 'Maladaptation and development as usual? Investigation climate change mitigation and adaptation projects in Cambodia', *Climate Policy*, vol. 19, no. 51, pp. 547–562.

Zalewski, M 2010, '"I don't even know what gender is": a discussion of the connections between gender, gender mainstreaming and feminist theory', *Review of International Studies*, vol. 36, no. 1, pp. 3–27.

2 Methodology
Critical, conceptual and empirical issues

Introduction

The following pages provide an overview of how the research has been operationalised through a dual research methodological framework, inclusive of conceptual research, and a small empirical study undertaken in Indonesia in 2014–2015. The term conceptual research is used interchangeably with critical theoretical research, whose methods include critiques of dominant ideologies and research orientations, paradoxes, contradictions, assumptions and, more broadly, political ideological ideas, for example, neoliberal market environmentalism, to address the key question of this book: 'What is the basis of just adaptation and mitigation policies that can be adopted by the United Nations Framework Convention on Climate Change (UNFCCC) once these have been subsumed within a global carbon market framework?'

Addressing this question analytically required the broad resources of a multi-disciplinary critical theoretical approach to reflect the myriad ways in which climate change intersects with human-nature relations, exemplified in the notion of anthropogenic climate change. Additionally, the use of a range of inter- and intra-disciplinary theories, from the sciences (critical geography), quasi-sciences (neoliberal economics) and humanities (for example, feminisms), is not meant to overwhelm the reader, but to introduce how a range of different theories, from various disciplines, each in their own way contribute to the central question through a layering process. For example, the discussion between critical theory versus positivism in this chapter is crucial because the latter orientation has been historically privileged in climate change discourse, while the former has been neglected. Similarly, variations on adaptation approaches and types in the following chapter bring forth a new conceptual development – that is, the notion of 'precautious adaptation' emerging from an existing critique of adaptation types, that does not adequately capture this type of adaptation. The idea of precautious adaptation inheres the avoidance of maladaptation, where Magnan et al. (2016, p. 646) advocate 'the need for putting the risk of maladaptation at the top of the planning agenda'. Similarly, in Chapter 5 intra-disciplinary differences between feminisms, such as ecofeminism and classical liberal feminism are crucial to understanding why women

Methodology 17

in the context of climate change have been side-lined, through a critical examination of the 'silence' surrounding women in climate change discourse. In essence, the gravity and urgency of the climate change problem necessitates a theoretically variegated approach, and a siloed disciplinary approach to investigating climate change solutions would be intellectually disingenuous, indeed irrelevant. Climate change and its solutions are messy, and this chapter presents a guide on how the book's research has been operationalised.

While a conceptual research approach works well enough, up to and including Chapter 5, Chapter 6 introduces *in situ* research in the form of empirical feminist qualitative research to infuse the book's methodological rigor, underpinned by the 'critical analytic framework' (M Wearing, pers. comm, 2016, 3 May).

Conceptual research using critical theoretical methodology

Research design

A conceptual research approach falls within a critical social science philosophical orientation. While this approach underpins the entire book, including a critical analysis of the empirical findings from the Indonesian study, it was necessary to weight this methodological approach more heavily in the first half of the book in the chapters on adaptation (Chapter 3), mitigation (Chapter 4) and REDD+ mitigation/adaptation to women (Chapter 5). The reason for this is because together they form the basis, the theoretical underbelly for understanding, on a deeper level, what emerged in relation to the findings from the on-the-ground empirical study that are presented in Chapters 6 and 7. The findings of the study, particularly with respect to women, when assessed in juxtaposition with conceptualisations such as women as 'silent offsets' in REDD+, more broadly the 'gendered silent offset economy' under conditions of carbon capitalism, makes this clear. In other words, discerning theoretically what a 'silent offset' means was buttressed from on-the-ground findings, particularly how the concept of 'gender mainstreaming' has played its part in the silence of women in the climate change and economic discourses this chapter illuminates.

The purpose of critical research has been variously described as: seeking 'to change the world' through conducting research that critiques and transforms social relations (Neuman & Kreuger 2003, p. 81); in a milder variation, helping to understand and change social reality (Lather 1992, cited in Sarantakos 1998, p. 39); and revealing myths and disclosing illusions. The aim of this type of research is to contribute to empowering and emancipating people in its milder variation, whereas social reality for the critical theorist is based on oppression and exploitation (Sarantakos 1998). Thus, critical theorists generally conduct research from an emancipatory interest that digs beneath surface meanings, or what Baldwin and Bettini (2017, p. 7) refer to as the 'political subsurface', of dominant institutional frameworks and methods of knowledge that reveal and

18 *Methodology*

endeavour to counter existing power relations, oppression, exploitation and marginalisation in all forms.

While an emancipatory thrust is an important one in critical theory, other motivating cognitive interests can also be important, such as a technical interest (Alvesson & Sköldberg 2009). An overriding motivational interest in this book is to suggest alternative ways of constructing climate change policies to avoid maladaptation. To this end, and in many respects, the emancipatory and technical thrust critical theorists adopt in their research is synonymous with that of a feminist standpoint theorist. That is, insofar as both seek to 'place relations between political and social power and knowledge centre-stage' (Bowell n.d.).

In the chapters on adaptation and mitigation that follow, the book is premised on both technical and emancipatory cognitive interests in critiquing mitigation and adaptation policies to climate change. In these chapters, I analyse the technical nature of climate change solutions, particularly mitigation policy, or the carbon offset mechanism of the Clean Development Mechanism (CDM) under the Kyoto Protocol. Similarly, adaptation policy is viewed within the dominant paradigm of 'development-as-usual' (Work et al. 2019). What is argued and developed through these critiques is that the solutions to climate change indicate maladaptation emerges through several pathways. By drawing attention to the maladaptive features of adaptation and mitigation policies, alternatives to the dominant technical and market-centric aspects currently present in both emerge. That is, and in particular, the carbon-centric focus that is accompanied by technical measurements, verification, monitoring and reporting essential to the commodification of carbon to the detriment and exclusion of alternatives that would be more transparent and inclusive of the communities where the carbon programmes are situated.

What also emerged from a technical critique of climate change solutions, was a salient invisibility and silence of women in the climate change discourse in general and climate change solutions in particular, which have been sidelined by both scientific and market epistemologies. Thus, Chapter 5 is a feminist critique of the existing dominant theories that explain how and why a 'gender gap' has emerged in the area of climate change. The silence around women is due to the privileging of productive economy. That is, where market-based solutions to climate change mitigation policy reside, over social reproduction where women and, as importantly, adaptation policies are situated. This, in, by and of itself, is highly maladaptive because as one economist rightly cautioned, 'mitigate we might; adapt we must' (Nordhaus 1994, p. 189, cited in Pielke 1998, p. 160).

The issue of climate change is then a problem that requires the full resources of trans-disciplinary critical research. This requires employing the theoretical resources of disciplines that incorporate both the study of the social sciences and natural sciences, such as critical geography's scholarly literature on 'neoliberal natures' (Bigger et al. 2018), in particular used in the structural critiques of climate change solutions. In 2009, when research for this book began life as a PhD thesis, it became obvious when reviewing the vast literature on the topic that climate change was regarded first and foremost as a science problem that could be addressed through technological and scientific means; hence utilising a critical research path emphasised a technical interest in my analysis over an emancipatory cognitive interest, though both are present.

The establishment of the Intergovernmental Panel on Climate Change (IPCC), the United Nations scientific body charged with providing Assessment Reports at regular intervals on the state of knowledge on climate change, confirmed the pre-eminence of science's role in addressing climate change. The global policy position that emerged in response to these periodic scientific reports has been identified in the literature as the 'mitigation bias' (Schipper 2006), where adaptation to climate change took second place. Indeed, until recently, to speak of adaptation was 'taboo' and tantamount to a defeatist position as discussed in the preceding chapter. Specifically, the Kyoto Protocol, which guides how climate change should be addressed, emphasised several market-based mechanisms, such as emissions trading and carbon offsets such as the CDM. However, in spite of the vast resources and time expended under this mitigation bias, the results have indicated that there have been increased global emissions in the first commitment period under the Kyoto Protocol (2008–2012) (Liverman 2009), and in its second commitment period (2013–2020) (Cames et al. 2016; Evans & Gabbatiss 2019a, 2019b).

Thus the positivist orientation in which climate change mitigation is located, with its emphasis on technical measurements to establish baselines, 'additionality', measurement, reporting and verification in order to quantify the amount of carbon reduced and then 'traded' or 'offset', is not only a scientific issue but a political one. That is, it was a political decision to emphasise mitigation over adaptation, in spite of both being given equal footing in the UNFCCC. It was also a political decision to use market-based mechanisms to give primacy to mitigation through the Conference of Parties (COP) who meet annually in negotiating the policy responses to the scientific reports emerging from the IPCC. Thus, the political-ideological dimension of critical theory is emphasised in this critique, where the politics of climate change solutions is necessarily placed within a 'setting for the continual formation of reflective political-ethical argumentations' (Alvesson & Sköldberg 2009, p. 147). The political-ethical argument for just solutions to climate change is most saliently addressed in Chapter 8 on justice. The normative concept of justice inheres ethical arguments in its constitution. In other words, existing climate change solutions present as unjust when they ignore indigenous knowledge and ways of protecting their environments, or where women are used instrumentally to achieve outcomes for predetermined global mitigation projects that do not take account of their social roles and subjective needs.

The use of critical theory flags a fundamental refusal to treat problems as discrete phenomena that can be manipulated with 'a bit of social [or scientific] engineering; rather they are viewed in light of the totality-subjectivity combination, that is, critical theory sees society in terms of culturally shared forms of

20   *Methodology*

consciousness and communication' (Alvesson & Sköldberg 2009, p. 159). A positivist paradigm does not, and cannot, theoretically view reality in this way. Even though positivism and a critical perspective conceive social reality as being 'out there' to be discovered, which falls within a realist position – where the two approaches differ is in the way critical theory views social reality as evolving over time. That is, reality is not stable but fluid. Positivism, on the other hand, views social reality as stable pre-existing patterns or order (Neuman & Kreuger 2003).

However, the role of human agency in climate change – that is, the recognition of anthropogenic climate change – should have put 'the final nail in the coffin of environmental determinism' (Head 2010, p. 236). That is, if 'humans and their activities are embedded in the very structure of the atmosphere, we needed new ways of thinking about things' (Head 2010, p. 236). With these thoughts in mind, critical theory was deemed the most appropriate orientation, with its variety of methods and the employment of a multi-disciplinary theoretical lens, as an attempt to find and develop new ways of thinking about climate change solutions. In particular, of what constitutes a just policy regime, addressed in the final chapter on 'Justice in the age of the Anthropocene'.

Methods of data collection and data analysis

A variety of methods are used in a critical theoretical approach, including analysis, reflection, critique and critical comparison (McMurray, Hall-Taylor & Porter 2002, p. 18). These methods have been applied to directly addressing the central question under analysis and, in particular, through an important method which involves the meticulous examination of existing theories (Alvesson & Sköldberg 2009). For example, alternative theories in the sciences and the humanities are used to critique the dominant theories of neoliberalism and global environmental economism that currently frame and shape unjust solutions under the UNFCCC to climate change. These various theories draw from: critical geography (Bigger et al. 2018; Bumpus & Liverman 2008; Castree 2006; Liverman 2009, 2011; McAfee 1999, 2014; McCarthy & Prudham 2004; Silvey 2009); ecological economics and critical accounting (Brown 2013; Daly 1995, 2005, 2008; Daly & Townsend 1993; Kosoy & Corbera 2010; Muradian et al. 2010). Additionally, in Chapter 5, feminist theories such as eco-feminism (Henderson 1982; Mellor 1996, 2012; Mies 2007) that directly critique status quo policy solutions to climate change, through the CDM and United Nations Programme on Reducing Emissions from Deforestation and Forest Degradation (UN-REDD+) Programme. The latter is a global carbon mitigation programme developed in collaboration with three United Nations agencies – the Food and Agriculture Organization (FAO), United Nations Environment Programme (UNEP) and United Nations Development Programme (UNDP). The essential idea behind REDD+ is to incentivise developing countries to reduce deforestation and forest degradation.

Methodology 21

What these various disciplinary theories hold in common, is a deep reservation that the dominant ideology, discourse, theory and practice of neoliberal environmentalism in general, and the market-based solution established under the Kyoto Protocol in particular, indeed the new mitigation mechanism established under Article 6.4 of the Paris Agreement, can 'solve' climate change. In this way, a well thought out theoretical frame of reference can facilitate making good interpretations, 'something which requires particular attention to in critical theory as the idea is to go beyond surface meanings' (Alvesson & Sköldberg 2009, p. 166). The following provides, by way of point, an example of how feminist theories are used to critique neoliberal environmentalism. In Chapter 5, a false universalism underscores current international adaptation and mitigation efforts based on a market-based epistemology; ecofeminism rejects neoliberal environmentalism for, among other aspects,

> ignoring the context within which environmental relations are nested – for example, women relate to natural resources as part of their livelihood strategies, which reflect multiple objectives, powerful wider political forces and, crucially, gender relations, i.e. social relations which systematically differentiate men and women in processes of production and reproduction. Micro-studies of resource use reveal that the relations of women to environments 'cannot be understood outside the context of gender relations in resource management and use'.
>
> (Leach 1991, p. 14, cited in Jackson 1993, p. 1949)

In presenting the outcomes of a small-scale empirical qualitative research study (the second methodology employed in the book's dual methodological framework), conducted in Indonesia in 2014–2015 focusing on the impacts of UN-REDD+ on 'adat' communities, women specifically, what the *in situ* study revealed was the primacy of 'livelihood strategies' in the thinking of women, and not carbon mitigation as REDD+ is currently framed. Indeed, research conducted so far on engendering climate change responses in the secondary sources, while no doubt well meaning, has focused endlessly on gender mainstreaming. The existing framing of gender, however, tends to accord and be congruent with a well-worn historical path of women of the South being used for otherwise instrumental ends to meet and serve strategic capital interests of the North. In other words, there is little evidence of a relational perspective to be found here except, that is, to the galloping juggernaut of the global carbon market where, for example, a business case for women's inclusion in this mythical creation is led by the United Nations in a paper entitled 'The business case for mainstreaming gender in REDD+' (UN-REDD Programme 2011).

Thus, not all feminist theories are on the same page, and the use of early ecofeminist theory to critique and re-evaluate the liberal feminist ideal of 'gender mainstreaming', exemplified in the UN-REDD 'business case' (UN-REDD Programme 2011), is a salient example. Classic liberal feminism (Bessis 2003), like all feminist theories in general, challenges gender bias and gender blindness.

22 *Methodology*

However, what the critical analytic method brings to the surface in its application to feminist theories, is that the liberal feminist ideal of women's 'equality' has reinforced the dominant, neoliberal approach to climate change solutions such as REDD+. Indeed, the authors of the 'business case for mainstreaming gender in REDD+' appear to legitimate the myopic status quo of the primacy of a neoliberal economic approach to procedural inclusion via gender mainstreaming. Further, this is particularly evident in their circumscription of sustainability of mitigation benefits as, in the end, the preservation of investor confidence through the minimisation of risk and project reversal. In fact, the business case says as much in the final paragraph of its 'Conclusions and Recommendations':

> Overall, giving consideration to gender equality in each readiness component of REDD+ makes good business sense, both creating and benefiting from a more stable investment environment for forest carbon assets.
> (UN-REDD Programme 2011, p. 7)

Whether or not the authors of the business case are aware of their own research and gender blindness in reinforcing the notion that a market-based mitigation solution is *the* solution to climate change, thus legitimising the dominant neoliberal environmentalist theory that underscores REDD+ is unclear. What is clear, though, and feminists observe is

> that research in general has been male-dominated, which has largely determined the questions addressed (or not addressed), and the way these have been answered ... In this way the asymmetrical gender relations in society are legitimated and reproduced.
> (Alvesson & Sköldberg 2009, p. 237)

Thus, a reflective research process central to a critical perspective must include self-awareness to mitigate dominant modes of thought being regarded as rational, natural and neutral (Alvesson & Sköldberg 2009, p. 176).

The purpose of the in-country study in Indonesia was to generate raw data expressing alternative modes of thought by those impacted on the ground through the mitigation programme of REDD+. Digging beneath the dominant strategy of gender equality employed by REDD+ – gender mainstreaming – there was an obvious disconnect between what is considered by gender theory, and project developers, to be a well-intentioned strategy, with its practice *in situ*. The term gender, let alone gender mainstreaming, was little understood by the participants, let alone the adat women and communities they represent. Further, it has failed to address the critical issues of improving their livelihoods, specifically in the area of health for their families and income-generating projects (beyond propagating seedlings and planting trees) that suit their subjective positions and needs. The insights gained from face-to-face interviews are thus crucial to reformulating mitigation projects that both eschew elite language and for project developers to address the specific needs the women have expressed themselves, when directly asked. As one participant observed, being sensitive to gender, taken to mean women, in the climate change negotiations does not mean they have been mainstreamed into the heart of the project itself (AP1 Jakarta 2014). That would require minimising elite language, learning to listen, and acting upon their specific expressed needs.

It has been necessary to present, by way of example above, how methods drawn from a critical theoretical orientation are employed and operationalised throughout the book. However, in addition to going beyond 'surface meanings', critical theory uses a praxis approach, or is action-oriented. That is, in an effort to distinguish between what is good theory and what is bad theory, a critical approach 'puts the theory into practice and uses the outcome of applications to reformulate theory' (Neuman & Kreuger 2003, p. 85). Indeed, the basic strategy of writing as resistance and activism is itself praxis. The voices and experiences of adat women and men gained in practice are used to reformulate aspects of current adaptation and mitigation policies they are subjected to.

Following the deeply complex and abstract account on the silence surrounding women in the climate change discourse, it became evident that a change in research orientation to empirical feminist qualitative research was needed to give voice to women (experts) in the climate change mitigation programme under UN-REDD+. This change in research orientation from the critical analytic framework was complemented by an empirical feminist qualitative research study, driven by a need to test theoretical insights regarding, for example, 'silent offsets', with on-the-ground accounts. While the conceptualisation of women as 'silent offsets' is to speak abstractly and theoretically, this was embedded in the empirical findings of the dedicated chapter 'Where are the women?'

The impacts of UN-REDD+ on indigenous women in Indonesia – grounded qualitative feminist research study

Research design: qualitative research: empirical qualitative feminist research

The Indonesian study was investigated within a qualitative methodological framework, grounded in 'feminist methodology in an effort to give voice to women, and to correct the male-oriented perspective that has predominated in the development of social science' (Neuman & Kreuger 2003, p. 88). The predomination of 'an overwhelming male-centric approach to the study of our social world' (Walter 2010, p. 21) has crucially determined what questions are addressed, as importantly those that are not, and in the way they have been answered. Thus, a feminist perspective brings to light through critique and re-evaluation of dominant theories, the gender bias and gender blindness which, in particular, [climate change] research has suffered (Alvesson & Sköldberg 2009). Feminist research thus differs from traditional research in several ways. The main focus centres on the viewpoints and experiences of women; there is a

24　*Methodology*

concerted effort to reduce or eliminate the power imbalance between researcher and respondents; and feminist researchers share an overall goal to change social inequality between women and men which renders or classifies a feminist research approach as action-oriented (Kumar 2014, p. 160).

A method of self-reflection is central to a qualitative feminist frame and, as the author, I had to challenge my own conceptualisation of women as 'silent offsets'. As well, inviting the voices of women in the study helped to challenge other feminist creeds – for example, 'gender mainstreaming' climate change programmes such as REDD+ – that are theoretically meant to ensure women are actively and representationally involved in the projects, but in practice held no resonance with women on the ground in the Indonesian study. By inviting the voices of women impacted by REDD+ in the study, this formed an important interconnection between theory and practice, by operationalising action-oriented participative research in the study. I listened, I heard, I revised and then reformulated earlier theoretical insights to reflect voices on the ground which has contributed to a more reflectively nuanced, rich and inclusive form of praxis.

Methods of data collection

Within a feminist research framework, semi-structured or unstructured interviewing are prominent methods of data gathering, where the 'in-depth face-to face interview has become the paradigmatic "feminist method"' (Kelly 1994, p. 34, cited in Bryman 2012, p. 491). The choice of using a semi-structured questionnaire as the main method of data collection was designed to ensure sufficient flexibility and responsiveness to the questions remaining open and mutual which, for example, a standard structured survey could not achieve. That is because a standard structured survey locks the responses in a uni-directional or one-way process, where the researcher extracts information from the interviewee (Oakley 1981, cited in Bryman 2012). As well, there is little flexibility in a structured survey, where the researcher cannot respond to questions that the respondent may ask in the course of the research process. Therefore, there is an element of hierarchy introduced by the use of the structured survey, where the power of the interviewer is revealed in seeking information from the respondent (Oakley 1981, cited in Bryman 2012, pp. 491–492).

Thus, in conducting the semi-structured questionnaire with the Indonesian respondents, the notion of a detached and objective researcher is epistemologically and methodologically incompatible with feminist qualitative research. Under a feminist approach, the objective is to collaborate and interact with the respondents in the data collection process, where the personal and the professional are fused so as 'to comprehend an interviewee's experiences while sharing their own feelings and experiences' (Neuman & Kreuger 2003, p. 88). The Indonesian study in its approach followed what feminist researchers advocate: one that establishes a high level of rapport between researcher and respondent; ensures a high degree of reciprocity as interviewer is maintained; develops a non-hierarchical relationship; and is interested in the perspective of the women being interviewed (Oakley 1981, cited in Bryman 2012, p. 492).

Taken together these formed a base for establishing a close and mutual relationship between the respondents and myself as researcher in the Indonesian study. With one crucial respondent, I spent two weeks in June 2014 in Central Kalimantan, and then returned again in December 2014 when I spent a further two months, informally discussing aspects of the research, living in the family home, and undertaking a site visit that was central to my research on REDD+. This type of personal relationship that matured over time, the blurring of the formal and personal relations in the research project, is characteristic of much feminist research, 'if not all' (Reinharz 1992, cited in Neuman & Kreuger 2003). Indeed, '"unless a relationship of trust is developed, we can have no confidence that our research on women's lives and consciousness accurately represents what is significant in their everyday lives, and thus has validity in that sense"' (Acker 1991, p. 149, cited in Alvesson & Sköldberg 2009, p. 242).

Conclusion

In summary, a dual research methodological approach using critical theory and qualitative feminist research has allowed for testing conceptual insights to have emerged through the critical analytic frame, with on-the-ground accounts gained in the qualitative study. This mixed approach has been significant in developing grounded theoretical insights about the issue of maladaptation in-country.

References

Alvesson, M & Sköldberg, K 2009, *Reflexive methodology: new vistas for qualitative research*, Sage, London.

Baldwin, A & Bettini, G (eds) 2017, *Adrift: climate change, migration, critique*, Rowman & Littlefield International, New York.

Barnett, J & O'Neill, S 2010, 'Maladaptation', *Global Environmental Change*, vol. 20, pp. 211–213.

Bessis, S 2003, 'International organisations and gender: new paradigms and old habits', *Signs: Journal of Women in Culture and Society*, vol. 29, no. 2, pp. 633–647.

Bigger, P, Dempsey, J, Asiyanbi, AP, Kay, K, Lave, R, Mansfield, B, Osborne, T, Robersson, M & Simon, GL 2018, 'Reflecting on neoliberal natures: an exchange', *Environment and Planning E: Nature and Space*, vol. 1, no. 1–2, pp. 25–75.

Bowell, T n.d., 'Feminist standpoint theory', *Internet Encyclopedia of Philosophy*, wiki article, n.d., accessed 30 June 2017, www.iep.utm.edu/fem-stan/.

Brown, L 2013, 'Herman Daly Festschrift: restructuring taxes to create an honest market', in J Farley (ed.), *The Encyclopedia of Earth*, wiki article, 7 October, accessed 17 June 2017, http://editors.eol.org/eoearth/wiki/Herman_Daly_Festschrift:_Restructuring_taxes_to_create_an_honest_market.

Bryman, A 2012, *Social research methods*, 2nd edn, Oxford University Press, Oxford.

Bumpus, A & Liverman, D 2008, 'Accumulation by decarbonisation and the governance of carbon offsets', *Economic Geography*, vol. 84, no. 2, pp. 127–155.

26 *Methodology*

Cames, M, Harthan, RO, Füssler, J, Lazarus, M, Lee, CM, Erickson, P & Spalding-Fecher, R 2016, *How additional is the Clean Development Mechanism?: analysis of the application of current tools and proposed alternatives*, Study prepared for DG CLIMA, Öko Institut, accessed 7 August 2020, https://ec.europa.eu/clima/sites/clima/files/ets/docs/clean_dev_mechanism_en.pdf.

Castree, N 2006, 'From neoliberalism to neoliberalisation: consolations, confusions, and necessary illusions', *Environment and Planning A*, vol. 38, no. 1, pp. 1–6.

Daly, H 1995, 'The irrationality of homo economicus', *Developing Ideas Digest*, Developing Ideas interview by Karl Hansen, College Park, Maryland, USA, 8 February 1995, accessed 3 June 2017, http://pratclif.com/sustainability/Herman%20Daly.htm.

Daly, H 2005, 'Living in a finite world: Herman Daly and economics', in CN McDaniel, *Wisdom for a livable planet: the visionary work of Terri Swearingen, Dave Foreman, Wes Jackson, Helena Norberg-Hodge, Werner Fornos, Herman Daly, Stephen Schneider, and David Orr*, Trinity University Press, San Antonio.

Daly, H 2008, 'Special report: economics blind spot is a disaster for the planet', *New Scientist*, vol. 200, no. 2678, pp. 46–47.

Daly, H & Townsend, K 1993, *Valuing the earth: economics, ecology, ethics*, MIT Press.

Evans, S & Gabbatiss, J 2019a, 'In-depth Q & A: how "Article 6" carbon markets could "make or break" the Paris Agreement', *Carbon Brief*, accessed 8 August 2020, www.carbonbrief.org/in-depth-q-and-a-how-article-6-carbon-markets-could-make-or-break-the-paris-agreement.

Evans, S & Gabbatiss, J 2019b, 'COP25: Key Outcomes agreed at the UN climate talks in Madrid', *Carbon Brief*, accessed 8 August 2020, www.carbonbrief.org/cop25-key-outcomes-agreed-at-the-un-climate-talks-in-madrid.

Head, L 2010, 'Cultural ecology: adaptation-retrofitting a concept?', *Progress in Human Geography*, vol. 34, no. 2, pp. 234–242.

Henderson, H 1982, *Total productive system of an industrial society: layer cake with icing*, digital image, accessed 14 May 2017, http://hazelhenderson.com/wp-content/uploads/totalProductiveSystemIndustrialSociety.jpg.

Jackson, C 1993, 'Doing what comes naturally? women and environment in development', *World Development*, vol. 21, no. 12, pp. 1947–1963.

Kosoy, N & Corbera, E 2010, 'Payments for ecosystem services as commodity fetishism', *Ecological Economics*, vol. 69, no. 6, pp. 1228–1236.

Kumar, R 2014, *Research methodology: a step-by-step guide for beginners*, 4th edn, Sage Publications, London.

Liverman, D 2009, 'Conventions of climate change: constructions of danger and the dispossession of the atmosphere', *Journal of Historical Geography*, vol. 35, no. 2, pp. 279–296.

Liverman, D 2011, *Amsterdam Global Climate Institute, No1–4.wmv*, uploaded 19 September 2011, online video, accessed 16 May 2017, www.youtube.com/watch?v=O5Nl0Gviz9s.

Magnan, AK, Schipper, ELF, Burkett, M, Bharwani, S, Burton, I, Erikson, S, Gemenne, F, Schaar, J & Zievogel, G 2016, 'Addressing the risk of maladaptation to climate change', *Wiley Interdisciplinary Reviews: Climate Change*, vol. 7, September/October, pp. 646–665.

McAfee, K 1999, 'Selling nature to save it? biodiversity and green developmentalism', *Environment and Planning D: Society and Space*, vol. 17, pp. 133–154.

McAfee, K 2014, *Green economy or buen vivir: can capitalism save itself*, Lunchtime Colloquium, Rachel Carson Centre for Environment and Society, online video, accessed 17 May 2020, www.carsoncenter.unimuenchen.de/events_conf_seminars/event_history/2014-events/2014_lc/index.html.

McCarthy, J & Prudham, S 2004, 'Neoliberal nature and the nature of neoliberalism', *Geoforum*, vol. 35, no. 3, pp. 275–283.

McMurray, D, Hall-Taylor, B & Porter, E 2002, *Unit EDU40001 research methods in social sciences: study guide*, School of Social and Workplace Development, Southern Cross University, Lismore, Australia.

Mellor, M 1996, 'Myths and realities: a reply to Cecile Jackson', *New Left Review*, no. 217, May/Jun, pp. 132–137.

Mellor, M 2012, *Just banking conference*, Friends of the Earth Scotland, online video, accessed 14 May 2017, www.youtube.com/watch?v=i-9vMXfTM10.

Mies, M 2007, 'Patriarchy and accumulation on a world scale revisited: keynote lecture at the Green Economics Institute, Reading, 29 October 2005', *International Journal of Green Economics*, vol. 1, no. 3–4, pp. 268–275.

Muradian, R, Corbera, E, Pascual, U, Kosoy, N & May, P 2010, 'Reconciling theory and practice: an alternative conceptual framework for understanding payments for environmental services', *Ecological Economics*, vol. 69, no. 6, pp. 1202–1208.

Neuman, WL & Kreuger, L 2003, *Social work research methods: qualitative and quantitative approaches*, 1st edn, Allyn and Bacon, Boston.

Pielke, R 1998, 'Rethinking the role of adaptation in climate policy', *Global Environmental Change*, vol. 8, no. 2, pp. 159–170.

Sarantakos S 1998, *Social Research*, 2nd edn, Macmillan Education Australia Pty Ltd, South Yarra, Australia.

Schipper, ELF 2006, 'Conceptual history of adaptation in the UNFCCC process', *Review of European Community & International Environmental Law*, vol. 15, no. 1, pp. 82–92.

Silvey, R 2009, 'Development and geography: anxious times, anaemic geographies, and migration', *Progress in Human Geography*, vol. 33, no. 4, pp. 507–515.

United Nations Framework Convention on Climate Change (UNFCCC) 1992, *United Nations Framework Convention on Climate Change*, United Nations General Assembly, 20 January 1994, accessed 9 May 2017, www.refworld.org/docid/3b00f2770.html.

United Nations Framework Convention on Climate Change (UNFCCC) 2015, *Paris Agreement*, accessed 16 September 2020, https://unfccc.int/process-and-meetings/the-paris-agreement/the-paris-agreement.

United Nations Programme on Reducing Emissions from Deforestation and Forest Degradation (UN-REDD Programme) 2011, *The business case for mainstreaming gender in REDD+*, The United Nations Collaborative Programme on Reducing Emissions from Deforestation and Forest Degradation in Developing Countries, accessed 15 May 2017, www.undp.org/content/dam/undp/library/gender/Gender%20and%20Environment/Low_Res_Bus_Case_Mainstreaming%20Gender_REDD+.pdf.

Walter, M 2010, *Social research methods: an Australian perspective*, 2nd edn, Oxford University Press, South Melbourne.

Work, C, Rong, V, Song, D & Scheidel, A 2019, 'Maladaptation and development as usual? Investigation climate change mitigation and adaptation projects in Cambodia', *Climate Policy*, vol. 19, no. 51, pp. 547–562.

3 Adaptation, development, maladaptation
Theory and practice

Introduction

There is an ever expanding literature on adaptation to climate change, enriched in particular by insights, empirical evidence and theoretical rigor from environmental geographers (Castree et al. 2009). Nonetheless, adaptation 'may be overwhelmed by its own popularity and all meaning slowly leaks out of it' (Burton 2009, p. 11). The confusion surrounding adaptation is due to its various interpretations emerging from the United Nations Framework Convention on Climate Change (UNFCCC), mainstream development and science communities. There has been an ongoing failure to evaluate climate change adaptation policies through the lens of maladaptation and this represents a serious oversight by the UNFCCC and mainstream development communities. This is understandable, given maladaptation is only now presenting as a growing field of research as the empirical experiences of adaptation policies have been accumulating rapidly (Juhola et al. 2016).

The temporal dimension of these accumulating adaptation experiences (Magnan et al. 2016; Juhola et al. 2016; Work et al. 2019) calls for a look at how adaptation has been framed, thus interpreted, at an earlier point in time. Eriksen et al. (2011) 'consider adaptation a process and further stress the temporal dimension as a crucial element to assess whether or not an initiative is, or could be, detrimental' (Erikesen et al. 2011, cited in Magnan et al. 2016, p. 649). The temporal dimension then is as important as the framing of adaptation. As the introduction to gender in the conceptual overview demonstrates, the scientific framing of climate change as gender neutral had profound implications for women's participation being leveraged to serve the strategic interests of carbon capitalism, which cares little for their subjective needs or socially allocated roles in deeply complex landscapes that the carbon projects are located in.

Overview of adaptation

The conceptual overview introduced and explained key terms and concepts as they are understood, used and linked throughout the book. This chapter explores the nexus between adaptation policy and maladaptation in particular,

Adaptation, development, maladaptation   29

and how the various interpretations of adaptation are leading to negative consequences in the form of maladaptation. The suggestion made here is that conceptual confusion over adaptation is occurring now with respect to international adaptation policy directed under the climate change community, made up largely of the Intergovernmental Panel on Climate Change (IPCC) and UNFCCC (Pielke 2009), together with the mainstream development community. Maladaptation emerges, in part, from adaptation policy being rendered meaningless under the weight of its own conceptual confusion. Prosecuting that argument requires backtracking to the marginalisation of adaptation policy under the 1992 UNFCCC which, historically and inherently, has a mitigation bias as discussed earlier.

In part, concepts drawn from Barnett and O'Neill's (2010) typology help to explain how maladaptation arises due to the UNFCCC's interpretation of adaptation. For example, Barnett and O'Neill (2010) provide examples of adaptation case studies describing maladaptation spanning geographical and temporal scales as well as in a range of sectors from agriculture, water management, health and infrastructure. However, few of the examples they reviewed for their seminal article on maladaptation, 'describe in any detail how maladaptive practices arise' (Barnett & O'Neill 2010, p. 211). It was to this gap in the literature the authors developed and systematised a typology of how maladaptation arises through five distinct pathways. Others have since followed, notably Juhola et al. (2016) and Magnan et al. (2016), both of whom have further refined maladaptation by analytically sharpening and strengthening the concept to operationalise its use in practice. Thus, maladaptation's explanatory power lies in its guiding ability to screen various interpretations of adaptation policy to avoid maladaptive outcomes, such as 'precautious adaptation' which is advanced in the chapter and elaborated upon later.

The literature on adaptation policy to climate change broadly identifies different 'types' or interpretations of adaptation from two distinct communities – on the one hand is the climate change community and on the other, the development community (Ayers & Dodman 2010). This community binary's broad philosophical roots are found, respectively, in the divide between the natural and physical sciences, and the social sciences and humanities traditions. The discipline of geography's theoretical utility lies directly with its own explanatory power, that is to

> 'bridge one of the greatest of all gaps': namely, that 'separating the natural sciences and the study of humanity' ... [Thus] Geography remains one of the few disciplines committed to bridging the divide between the natural and physical sciences, on the one side, and the social sciences and humanities on the other – human-environment relations.
>
> (Mackinder 1887, cited in Castree et al. 2009, p. 1)

The broad literature of environmental geography, and its various sub-fields of economic geography, political, human, cultural and political ecology

30 *Adaptation, development, maladaptation*

geographies, to name a few (Castree et al. 2009), are invaluable resources used in the critical analysis of adaptation, and mitigation in the following chapter. That is, environmental geography is made up, and provides a rich, informative and practically relevant range of work, ideas and insights crucial to critiquing current adaptation and mitigation policies, and the broad theoretic this literature brings to the analysis would be all the more impoverished by its exclusion.

A cursory look at the maladaptation typology developed by geographers Barnett and O'Neill (2010), demonstrates how geographers bring the social sciences and studies of humanity into their work, especially through a focus on the concept of 'vulnerability'. That is, one of the pathways through which maladaptation forms is when an adaptation action or policy disproportionately burdens the most vulnerable (Barnett & O'Neill 2010). Liverman (2011) urges geographers to overcome myths surrounding the social sciences and natural sciences, one of which is the persistence of environmental determinism. That is, environmental determinism assumes the environment is the most important factor influencing the impacts of disasters which has prevailed throughout the 20th century.

> Even now, much of the focus on explaining climate change and much of the focus on funding research is still on the climate and environmental side of things: reducing the uncertainty of modelling; improving the physical basis of crop models; predicting the paths of hurricanes. However, what social science has shown us over the past couple of decades, is that the impact of droughts and other climate events depend as much on social and economic conditions in a region as on a physical change in rainfall and other meteorological factors. This means to understand climate impacts you have to spend as much time, if not more, in understanding the patterns of poverty, technology and economy as you do analysing climate data or models. The term vulnerability is used to describe this idea that climate impacts are determined as much by society as they are by natural events ... we lack funding for large scale research projects as to how adaptation can reduce this vulnerability.
>
> (Liverman 2011)

This was Liverman's (2011) call to geographers to re-embrace human-environment relations that Mackinder rightly claimed is geography's distinguishing feature in bridging the natural sciences/social sciences binary. The breadth of environmental geography's theoretic is thus well suited to the task of critiquing adaptation policy emerging under the binary climate change and development communities for their potential for maladaptation.

What is saliently absent from the literature on the evolution of international adaptation policy, however, particularly in the IPCC and UNFCCC processes, is whether various adaptive strategies emerging from the different conceptions of adaptation have been screened for their potential side effects to avoid maladaptation. Scheraga and Grambsh (1998) suggest that maladaptation is one of the principles guiding effective adaptation policy.

Adaptation, development, maladaptation 31

Adaptation is mentioned five times in the UNFCCC text while maladaptation is not mentioned at all. Avoiding maladaptation, however, has a proxy found in the precautionary principle (PP), one of the principles guiding the UNFCCC's 'ultimate objective' of 'stabilisation of greenhouse gas concentrations in the atmosphere at a level that would prevent dangerous anthropogenic interference with the climate system' (UNFCCC 1992, Article 2). Article 3.3 of the UNFCCC begins with the statement that '[t]he Parties should take precautionary measures to anticipate, prevent or minimise the causes of climate change and mitigate its adverse effects' (UNFCCC 1992).

Screening adaptation policy for potential maladaptation is thus not only prudent and precautionary (Pielke 2009), but eschews the notion that precaution should be exclusive to mitigation as the Article's text suggests. Elliott (2007) says it is not clear that precaution only applies to a 'lack of full scientific certainty' – that is, lack of full scientific certainty should not be used as an excuse for inaction. Rather, Elliott (2007) argues precaution should be given standing as a more general principle due to the public goods nature of environmental protection and regulation.

This chapter supports the notion of generalising the principle of precaution in guiding adaptation policy, or what may be termed 'precautious adaptation' to avoid maladaptation. However, 'precautious' adaptation joins a list of many other adjectives differentiating adaptations, including autonomous, involuntary, passive, planned, anticipatory or reactive adaptation, among others (Eriksen et al. 2011). In addition, there have been calls for the promotion of a new approach called 'sustainable adaptation' in guiding climate change adaptation policy (Eriksen et al. 2011). That is, sustainable adaptation means 'adaptation measures that also contribute towards social equity and environmental integrity' (Brown 2011, p. 21). In other words, sustainable adaptation distinguishes itself from adaptation in general through qualifying actions to climate change in terms of their effects on social justice and environmental integrity. Thus, for any meaningful notion of sustainable adaptation policy to exist requires that these two features be met (Eriksen et al. 2011). If either feature is absent, or both, then the adaptation measure is rendered unsustainable. The question to emerge here is: what is the relationship between unsustainable adaptation and maladaptation?

Following Barnett and O'Neill's (2010) pathways to maladaptation, if an adaptation action or policy disproportionately burdens the most vulnerable, then this is one path through which maladaptation arises. If 'sustainable adaptation' qualifies adaptation actions in terms of their effects on social justice as one of two features that make for sustainable adaptation, then disproportionately burdening the vulnerable is socially unjust. Thus, the adaptation is rendered unsustainable, or maladaptive.

For now, what is ultimately lost in the binary discourse that characterises much of climate change solutions – mitigation/adaptation, whose epistemological roots are found respectively in the natural sciences/social sciences binary – is the higher order goal that the precautionary principle serves in

32 *Adaptation, development, maladaptation*

responding to anthropogenic climate change – the prevention of 'serious or irreversible damage' (UNFCCC 1992, Article 3.3). Reference to this higher order goal is not meant to imply, nor is it suggested, a return to simple environmental determinism because, as Head (2010) rightly acknowledges, in many ways recognition of anthropogenic climate change was 'the final nail in the coffin of environmental determinism' (Head 2010, p. 236). Additionally, she argues that if 'humans and their activities are embedded in the very structure of the atmosphere, we needed new ways of thinking about things' (Head 2010, p. 236).

In support of Head's (2010) call for new thinking, particularly in its application to climate change mitigation and adaptation solutions, this book adopts a new way of looking at mitigation policy – through the lens of maladaptation – to demonstrate how current international mitigation policy under the UNFCCC is on a maladaptive trajectory. The critique of mitigation policy through the lens of a multi-disciplinary critical theoretical analysis, is inclusive of environmental geography's sub-fields of economic and political-ecological geography as well as critical accounting theories among others, to counter the dominant ideological flourishes of neoliberalism's obsession with market-based mechanisms to mitigate greenhouse gas emissions. For example, through the Clean Development Mechanism (CDM) and Emissions Trading embedded in the universalism of the UNFCCC 1997 Kyoto Protocol. These so-called innovative market mechanisms simply exemplify old thinking that has resulted in the commodification of the atmosphere, the latest iteration of Marxian 'primitive accumulation' or what Harvey (2003) calls 'accumulation by dispossession'. Contemporaneously, and in the context of climate change, Bumpus and Liverman (2008) call this 'accumulation by decarbonisation'.

While a detailed description of what these terms mean is attended in the chapter on mitigation, briefly, they represent the theoretical backdrop to a discussion about the social and environmental consequences that accompany the monetisation of the atmosphere. With respect to adaptation, there is a close connection between financing adaptation through mitigation and the carbon market. That is, the Kyoto Protocol Adaptation Fund receives funding that is generated by a 2 per cent levy on projects developed under the CDM (Gender CC 2011; UNFCCC Adaptation Fund 2011; UNFCCC CDM 2011). The chapter questions the 'single language of valuation' and 'social emptiness' that characterise neoliberal solutions to climate change, and rejects their minimisation as mere irony. That is, ex-World Bank Chief Economist Sir Nicholas Stern argued that growth in the emission of greenhouse gases, in his 2006 Stern Report on Climate Change for the UK government, is a major market failure (Schalatek et al. 2010). However, the Stern Report then proceeded to advocate its solution in a global carbon cap and trade scheme (emissions trading) to be in place by 2020 to 'deal with the most threatening problem of our time' (Bello 2009, p. 5).[1] Others have, therefore, rightfully questioned the logic of market-based solutions in tackling climate change and how this should not, to any degree, 'be entrusted to the irrationalities and potentially speculative exuberance of an emboldened global carbon market'

Adaptation, development, maladaptation 33

(Schalatek et al. 2010, p. 5). Bond (2011) is right in arguing that since the 2008 financial crisis, financiers should not be allowed to bet on anything, let alone something as precious as the planet.

The concluding comment to mitigation policy is that the CDM market mechanism distinguishes itself as a mechanism for maladaptation and not mitigation. In developing that argument, the maladaptation typology is used to screen international mitigation policy and demonstrates how marketised mitigation not only does not meet the higher order goal of the Climate Convention's 'ultimate objective' of stabilisation of greenhouse gas emissions, but further breaches the already maligned PP as well.

This chapter employs the proxy of precaution or the PP to screen international adaptation policy for its potential to lead to forms of maladaptation.

The precautionary principle, like that of 'common but differentiated responsibilities', features in the 1992 Climate Convention (UNFCCC) as a principle guiding climate change responses. More generally, these key principles have been integral in guiding the development of international environmental law over the past 30 years. While both are important principles that 'stand out' as theoretically crucial, both are poorly implemented in practice (Elliott 2007, p. 11). The focus here is to apply the precautionary principle to adaptation policy. The focus on precaution overlaps with Barnett and O'Neill's (2010) pathways to maladaptation. That is, when adaptation policy is viewed in relation to the PP as a more general principle, maladaptation can potentially be avoided.

What is important to foreground now, is how maladaptation can emerge under different conceptions of adaptation, and how existing development models frame current approaches to adaptation (Brooks et al. 2009). Thus, the adaptation/development nexus is examined, where Brooks et al. (2009) claim that there is an urgent need in 'building development around the environment rather than vice versa' (2009, p. 758), as mainstream development theory and practice currently subscribe to. In other words, under the status quo, adaptation to climate change is perceived as something to be 'tacked onto' development rather than being integral to it (Ayers & Dodman 2010). It would appear old thinking is, indeed, one of the major impediments to avoiding potential maladaptation under different interpretations of adaptation, that is, mainstream development.

The central question under analysis in this chapter asks, are current conceptions of adaptation leading to maladaptation? Some interpretations of adaptation emerging from the development and climate change communities suggest this to be the case. A brief overview and introduction of how the chapter's sections unfold now follows.

Beginning in section one with the adaptation policy lag and the precautionary principle, this section recaps an overview of why adaptation has been slow to gain policy recognition unlike its cousin, mitigation policy, under the UNFCCC process. This apparent 'trade-off', or 'mitigation bias', that has prevailed under the UNFCCC and Kyoto Protocol took a decade before

34 *Adaptation, development, maladaptation*

adaptation policy, at the Conference of the Parties (COP) 13 in Bali in 2007, gained 'equal footing alongside mitigation', technology transfer and finance policies as four 'building blocks' required in response to climate change (Ayers & Dodman 2010, p. 163).

The second section describes how 'types or interpretations of adaptation' are better understood in relation to the evolving conceptual history of adaptation under the UNFCCC process which facilitates an understanding of the fragmentation of adaptation policy. In many respects this fragmentation stems from the lack of a definition of adaptation and specific provisions in the UNFCCC (Schipper 2006, 2009). In terms of understanding adaptation then, Ayers and Dodman (2010) see three 'types' of adaptation emerging from the development and climate change communities: from the UNFCCC who receives its scientific guidance from the IPCC, the interpretation of adaptation is 'stand-alone'; from the development community 'adaptation plus development' emerges, where climate-proofing development is the goal; and 'adaptation as development' from community-based adaptation (CBA) practitioners, where 'development is the basis for, and in some cases synonymous with, adaptation' (Ayers & Dodman 2010, p. 165).

The third and final section addresses whether current approaches to adaptation are leading to maladaptation. What conception/s or interpretation/s of adaptation policy should, therefore, be adopted or discarded to avoid maladaptation? This question presupposes there are existing problems in current adaptation policy that are producing maladaptive outcomes, and these details came to light in the preceding sections. For example, the UNFCCC mitigation bias created by the climate change community who favour natural science approaches to reducing greenhouse gas emissions (Huq et al. 2006) disproportionately dominated the policy space and agenda up until 2007 when adaptation finally gained equal footing with mitigation at COP 13 in Bali. 'In fact, during the first ten years the IPCC did not devote a single chapter of their reports to adaptation to climate change' (Klein n.d., cited in Hall 2016, p. 24). That is, a decade after mitigation had its very own Kyoto Protocol in 1997. The more relevant temporal implication of adaptation being considered the secondary response under the UNFCCC, is that the research and literature on mitigation and adaptation responses have advanced at different speeds, which, in turn, contributed to a lack of consensus and understanding about adaptation (Schipper 2006, 2009).

(1) International adaptation policy and the precautionary principle

Has adaptation become overwhelmed by its own popularity, such that all meaning has slowly leaked out of it as Burton (2009) predicted? If the previous overview to adaptation is any guide, Burton may, in part, be right about adaptation's 'popularity'; however, whether this extends to all meaning being leeched from it is up for review.

Adaptation's 'equal footing' alongside mitigation was a recent outcome of the UNFCCC process at COP 13 in Bali 2007 (Ayers & Dodman 2010). What

Adaptation, development, maladaptation 35

preceded adaptation's elevation at COP 13, however, dates back six years earlier to the publication of the IPCC's Third Assessment Report in 2001. According to Schipper (2006), shortly after this publication, a discussion on how to address the impacts of climate change resulted in 'an agenda item taking up adaptation [that] was introduced in the UNFCCC Subsidiary Body on Scientific and Technical Advice (SBSTA) ... and, in 2004, a work program on adaptation was adopted' (Schipper 2006, p. 368). The latter outcome, Schipper (2006) adds, was considered a 'considerable breakthrough'. This breakthrough culminated three years later at COP 13 in Bali in 2007 when adaptation policy, as suggested earlier, was placed on equal footing with mitigation, highlighting adaptation as one of four 'building blocks', alongside mitigation, finance and technical co-operation required in response to climate change (Ayers & Dodman 2010).

Schipper (2006) thus provides an important and rich historical overview of adaptation's evolution under the UNFCCC process; why mitigation policy developed as the primary response to addressing climate change, in spite of the IPCC identifying adaptation as the other. Reiterating aspects of Schipper's (2006) historical analysis then offers a partial explanation for how the policy bias emerged in mitigation's favour. That is, scientists working on adaptation issues in drafting the 1992 Climate Convention took adaptation to climate change to mean 'adaptive capacity'. That is, the limits to which a system can be stressed before it would collapse. Consequently, the emphasis on 'tolerable limits', more precisely 'ecological limits' or 'limits of tolerance', offers an explanation of why and how adaptation became a secondary consideration to mitigation, which reflected an initial application of the term adaptive capacity in assessing how much mitigation was needed. However, as Schipper (2006) reminds us, scientists drafting the Climate Convention construed adaptive capacity and system collapse in the context of both ecosystems and human systems, in line with the Club of Rome's 1972 report *The Limits to Growth*.

Adaptive capacity in the context of climate change is thus dependent on development contexts (Ayers & Dodman 2010), which must not breach limits that result in ecosystem and human system collapse. The application of the PP to climate change responses is thus, theoretically, meant to avert such extreme outcomes. The PP has a core meaning that applies to a particular category of new risks; these new risks develop slowly, are large scale and often with irreversible consequences (Whiteside 2006, p. 29). Climate change presents as such a risk and the periodic Assessment Reports issued every five years by the IPCC provide the evidence that human beings are responsible. Notwithstanding this, the dominant international policy approach has privileged mitigation policy as the preferred solution, manifest in the 1997 Kyoto Protocol.

However, market-based mitigation to reduce greenhouse gas emissions under the Protocol – for example, via Emissions Trading and the CDM – not only undermines the precautionary principle embodied in Article 3.3 of the Climate Convention but directly conflicts with its intent: 'The Parties should

36 *Adaptation, development, maladaptation*

take precautionary measures to anticipate, prevent or minimise the causes of climate change and mitigate its adverse effects' (Article 3.3, UNFCCC 1992). The conflict appears to have vanished in the jubilation surrounding a market-based solution to climate change which was locked into international environmental law through the Kyoto Protocol and perceived as a triumph for the neoliberal project of market environmentalism (Liverman 2009).

However, 'mitigation and adaptation must go hand in hand' (Prins & Rayner 2007, p. 975), and neoliberal solutions to mitigation policy briefly sketched in here necessarily implicate adaptation. The most salient example of this connection is through the Kyoto Protocol Adaptation Fund that is funded by a 2 per cent levy on CDM projects undertaken in developing countries by developed countries party to the Kyoto Protocol. There are serious ecological and social justice issues that surround the subject of 'carbon offsets', and while the following chapter details these specifically in relation to this mechanism's potential to create maladaptation, the point here is to emphasise more generally the consequences of neoliberal market environmentalism as 'the solution' to climate change, and how the dual features of 'sustainable adaptation' – that is, social justice and environmental integrity – are deeply interlinked and compromised by this approach.

An insight from John Berger, 'that it is now "space not time that hides the consequences from us"' (cited in Katz 2001, p. 1232), should give pause to those who believe carbon offsets and emissions trading are exempt.

Adaptation actions to climate change operate on a similar spatio-temporal dimension. For example, 'analysis shows that current adaptation interventions are often not sustainable; they may undermine long term resilience and may even result in "maladaptation"' (Brown 2011, p. 22). It is to this latter state that the precautionary principle should guide adaptation policy in avoiding such an outcome. That is why Elliott (2007) argues that the principle should have more general standing due to the public goods nature of environmental protection, where precaution as a general principle ensures that all forms of consumption and production are pursued to minimise environmental consequences. Unbridled over-consumption and production, of course, are antecedents to unsustainable development, which Huq et al. (2006) say is responsible for climate change to begin with.

Just as 'excavating the layers of process' (Katz 2001) facilitating adaptation's equal footing as a policy response to climate change alongside mitigation demonstrates time-lag in UNFCCC practice, there is another shift underway that speaks of a similar time-lag: that of maladaptation's stunted acknowledgement by the IPCC. Although not defined in the IPCC 2007 Fourth Assessment Report, maladaptation was previously defined in the 2001 Third Assessment Report as 'an adaptation that does not succeed in reducing vulnerability but it increases it instead' (IPCC 2001, p. 990, cited in Barnett & O'Neill 2010, p. 211). By the Fifth Assessment Report in 2014, numerous references were made, 'including an entire section dedicated to "Addressing Maladaptation"' (Noble et al. 2014, cited in Magnan et al. 2016, p. 648).

Adaptation, development, maladaptation 37

Qualifying adaptation actions in terms of their effects on environmental integrity and social justice is what distinguishes 'sustainable adaptation' from adaptation in general, as previously suggested by Eriksen et al. (2011). 'Adaptation to what' is interpreted specifically to climate change within the international climate change framework of the UNFCCC rather than climate variability, including climate change, more broadly (Ayers & Dodman 2010). Further, these authors argue the consequences of 'adaptation as an issue of climate change has created policy frameworks that do not fit with defining adaptation in terms of sustainable development' (Ayers & Dodman 2010, p. 165).

Mitigation policy under the 1997 Kyoto Protocol makes provision for the concept of sustainable development in CDM offset projects. The analysis of whether the CDM meets the dual objective of sustainable development and 'additionality' (that is, reductions in greenhouse gas emissions that are not business-as-usual) is grim in its conclusion: the CDM appears to distinguish itself more as a mechanism for maladaptation than mitigation. The reason for this conclusion centres on 'sustainable development', crudely and reductively, as having no monetary value in the [CDM] mechanism (Paulsson 2009, p. 70). The term is further hindered by the lack of an operational definition (Boyd et al. 2009; Brown 2011). Therefore, the language of trade-offs, social and ecological externalities, or the negative consequences of responses to climate change is invoked, almost normalised under this market-based mechanism in privileging spatially hidden cost-effectiveness.

In summary, these outcomes of mitigation policy dovetail adaptation responses to climate change thus giving rise to the new concept of sustainable adaptation (Eriksen et al. 2011), as previously suggested. However, as also suggested, Brown (2011) questions whether sustainable adaptation is nothing more than an oxymoron; that sustainable adaptation shares some similar features of sustainable development, a term she labels as 'deliberately vague and slippery' therefore making it difficult to operationalise; and that it fosters, among other things, the notion of sustainable growth. Brown (2011) calls out the slipperiness of 'sustainable adaptation', because unless sustainable adaptation specifically deals with fundamental problems in the dominant neoliberal paradigm of unsustainable development then it will remain an oxymoron. Indeed, what is needed is not further growth, as Lovelock (2006, 2009) suggests and the science on climate change demands, but a 'sustainable retreat' which necessarily entails a 'degrowth of the economy' (Brown 2011, p. 29).

However, the idea of degrowth conflicts with the notion of human progress, central to the philosophical enlightenment project, where the notion of 'retreat' does not mingle well. Indeed, the idea of progress, according to Gray (2007), is a secular version of the Christian belief in providence. Progress, however, has another source, 'in science, [where] the growth of knowledge is cumulative' (2007, p. xi). Outside of science, Gray (2007) argues, the idea of progress is simply a myth. The obvious question that Gray's assessment provokes is, how does one retreat from mythology? More pointedly, how does one retreat from neoliberal market environmentalism's stranglehold on the international

38 *Adaptation, development, maladaptation*

climate change negotiations and frameworks that do not seek a retreat from market-based solutions to climate change, but their further entrenchment?

At this point in the chapter, adaptation to climate change is perhaps not so much overwhelmed by its own popularity such that all meaning has slowly leaked from it; rather, the term is overburdened by a legacy of actual and philosophical conflations that begin with the adaptation/development policy nexus which is more deeply rooted in another, progress and myth, who conspire against any notion of retreat. This legacy has added to the conceptual confusion that Schipper (2006) says adaptation policy has experienced under the UNFCCC process. What, however, is less confusing, is the unfashionable root causes that have been illuminated in this brief analysis of what (un)sustainable adaptation/development means. In the context of climate change, what is ultimately required 'is to challenge and re-configure ideas about development' (Brown 2011, p. 29). On the one hand, the legacy of the dominant development paradigm is based on ideas of progress that are underpinned by processes of growth, modernisation and globalisation (Brooks et al. 2009). On the other, and for the first time Whiteside (2006) suggests, progress consists of humans recognising our inability to master the world. A key ethical expression of that recognition, she argues, is found in the precautionary principle.

The precautionary principle – 'precautious adaptation'

The key question the book addresses is 'what is the basis of just adaptation and mitigation policies that can be adopted by the UNFCCC once these have been subsumed within a global carbon market framework?' This question derives from Lohmann's (2008) broader question of what the project for a just solution to the climate crisis becomes, 'once it is associated with or incorporated in an economic development or carbon market framework?' (Lohmann 2008, p. 364). The answers to these questions are related to a larger question that remains unanswered. That is, 'whether the precautionary principle *necessarily* modifies the liberal philosophy embodied in existing international trade regimes' (Whiteside 2006, p. xii). This latter question goes to the heart of what is called 'precautious adaptation', thus eschewing the tired and tainted term 'sustainable' as a modifying adjective to climate change adaptation/development. That conflation is, above all, rooted in a paradigm that promotes economic growth, a major tenet of modern-day progress manifest, according to the degrowth advocates, in 'GDP, the fetish index of modern wellbeing' (Latouche 2010, p. 520).

John Gray (2007) may be right in his assessment that liberal societies, liberal humanism, the secular believers in particular, are in 'the grip of unexamined dogmas', such as neoliberal market environmentalism. Because, unlike science, where progress in the growth of knowledge is cumulative, this is not the experience of human life as a whole where history reminds us, time and again, that 'what is gained in one generation may be lost in the next' (Gray 2007,

p. xiii). Once again, the temporal dimension to climate change solutions are brought to the fore. That is, today's adaptation and mitigation policies may be tomorrow's maladaptation.

The insights of Berger that space, not time, now hides the consequences from us, and those of Gray's on time as social motion and dislocation, of loss and of gain intergenerationally, invokes time-lag in adaptation actions.

The literature on adaptation policy to climate change broadly identifies different 'types' of adaptation from the climate change and development communities (Ayers & Dodman 2010). The binary is philosophically reinforced and found respectively in the divide between the natural and physical sciences, and the social sciences and humanities traditions. Geography is one of the few disciplines that remain committed to bridging the gap separating 'the natural sciences and the study of humanity' (Mackinder 1887, cited in Castree et al. 2009, p. 1). The theoretical resources of critical geographers are used to navigate the discussion on adaptation policies, their insights of space and time, and of slow feedbacks that are spatially distant (Brown 2011), and among others, gauge whether various interpretations of current adaptation policies are leading to maladaptation.

(2) Types of adaptation – stand-alone, adaptation plus development, adaptation as development or community-based adaptation

This section describes different types of adaptation and suggests that community-based adaptation is closely allied with Brooks et al. (2009) in their call for building development around the environment instead of vice versa, as the case is with 'stand-alone' and 'adaptation plus development' interpretations. Recalling that environmental integrity is one of two features, along with social justice, that must be met to qualify adaptation actions as 'sustainable', Brown (2011) nonetheless questions whether sustainable adaptation is an internally contradictory concept, insofar as sustainable adaptation shares some similar features of sustainable development, that fosters, among other things, the notion of sustainable growth. Brown's (2011) assessment is that sustainable development and sustainable adaptation both share the same flawed feature of leaving the notion of 'sustainable growth' untouched.

If anthropogenic climate change is the result of unsustainable development (Huq et al. 2006), then for the pillar of environmental integrity to be preserved under sustainable adaptation, development adaptation theorists and practitioners must be prepared to embrace what some are calling for, '"a-growth" as one speaks about atheism … [which] actually means quite precisely the abandonment of a religion: the religion of the economy, growth, progress and development' (Latouche 2010, p. 519). Ecological economists have long been advocating the notion of a steady-state economy, or zero growth, 'a-growth', 'degrowth' or 'sustainable degrowth', and are under no illusion about the finiteness of the planet's resources or of how adapting to a climate-changed world requires its radical realisation (Kallis 2011). Not only does

40 *Adaptation, development, maladaptation*

environmental integrity depend on it, but social justice demands it: the two pillars of sustainable adaptation (Eriksen et al. 2011). Based on the foregoing, if another adjective is to be added to the term adaptation, then 'asymmetrical' appears a likely fit – a term that captures the notion that 'one group's adaptation is another group's hazard' (Kates 2009, p. 292).

As previously stated, the 'taboo on adaptation' is a policy outcome of the 'ecological approach' adopted by scientists working on adaptation issues in drafting the UNFCCC (Schipper 2006). Burton (2009) calls the ecological approach the 'pollutionist' view of adaptation, as it focuses on limits or levels of pollution of greenhouse gases that can be tolerated in the atmosphere. This view was what was 'uppermost in the minds of those who drafted the UNFCCC' (Burton 2009, p. 89). Climate change was thus defined as a 'science' problem, not a social one (Huq et al. 2006). However, as Schipper's (2006) history of adaptation in the UNFCCC process reminds us, the scientists who were working on adaptation issues in drafting the UNFCCC also construed adaptation as adaptive capacity in the context not just of ecosystems, but of human systems.

From adaptation's origins in the ecological/'pollutionist' view, various types of adaptation policy have emerged where some have reinforced the subordination of human systems to ecosystem collapse. Ayers and Dodman (2010) typify adaptation policy as broadly fragmented into three broad types under the climate change and development communities – 'stand-alone' adaptation from the UNFCCC, 'adaptation plus development' and 'adaptation as development'. The authors caution, however, against going too far and seeing each type as discrete. That is, 'not all adaptation is development, and not all development contributes to adaptation. Long-term adaptation priorities [to climate change] may conflict with near-term development priorities' (Ayers & Dodman 2010, p. 165). For example, the imperative of governments to address inequality and poverty in their own countries are political decisions that are framed in the short to medium term, and often framed against political cycles of incumbency. Whereas the long-term horizon trajectory that climate change adaptation measures require may conflict with these immediate imperatives. This is why adaptation as development, described below, can potentially bridge both horizons. In this way 'precautious adaptation' is observed.

Huq et al. (2006) suggest that one of the main reasons for the separate evolution of development policy and climate change policy, is attributable to the different scales, both geographic and temporal, at which problems are perceived (Huq et al. 2006). For example, relative to climate change discourse which is based on long-term projections of up to 100 years, according to development practitioners, this does not compare to the urgent needs of development concerns such as food security, or pollution or HIV/AIDS. Thus, the development scenarios are shorter term.

'Stand-alone' adaptation emerges from the UNFCCC. That is, the UNFCCC perceives adaptation narrowly as adaptation to climate change and not the broader notion of climate variability, which includes climate change as well. Accordingly, adaptation actions are limited to changes that are proven to be anthropogenic. 'Adaptation interventions are therefore "stand-alone" and "additional" to baseline development needs' (Ayers & Dodman 2010, p. 165).

The UNFCCC interpretation of adaptation is adopted by the Global Environmental Facility (GEF), which functions as the financial mechanism for the UNFCCC (Okereke 2007). The GEF manages the Least Developed Countries Fund and was established by the UNFCCC to address the adaptation needs of the 48 Least Developed Countries (LDCs), which are especially vulnerable to the adverse impacts of climate change. Activities supported under the LDCF include the preparation and implementation of National Adaptation Programmes of Action (NAPAs). NAPAs identify urgent and immediate adaptation needs and are not intended to address medium to long-term adaptation (GEF Secretariat 2009).

Ayers and Dodman (2010) highlight how problematic the UNFCCC interpretation of 'stand-alone' adaptation is in terms of LDCF finance for projects identified under the NAPAs. They cite the small island state of Tuvalu which identified one of their projects under their NAPA as

> coastal infrastructure to protect the shoreline from erosion, a problem regardless of climate change (and so a development need), but one exacerbated by climate change (so also an additional cost). The LDC Fund will only fund the additional cost of adaptation ... being a poor country, Tuvalu cannot afford to meet the costs of the baseline infrastructure. Thus the offer to fund, as it were, the 'top section' of the infrastructure required to respond to 'additional' impacts of climate change is absurd in light of the fact that co-financing to pay for the lower section cannot be found. The project is currently in limbo while co-financing is sought.
> (Ayers & Huq 2009, cited in Ayers & Dodman 2010, p. 166)

The language of 'climate proofing' development investments and 'mainstreaming' adaptation into development is employed as the solution to shore-up maintenance of the dominant development paradigm (Brooks et al. 2009). Within the adaptation framework offered by Ayers and Dodman (2010), this is typified as 'adaptation plus development'. That is, rather than building development around the environment, which is what is urgently needed, this approach builds the environment around development, thus ensuring the dominant paradigm and its central articles of faith remain intact. That is, the flawed dogmas of economic growth, modernity and progress driving current development approaches. Brooks et al. (2009) further argue that this approach must be challenged, indeed discarded.

Unlike 'adaptation plus development' which appears rooted in the developmentalist trap of universalism (Brooks et al. 2009), 'adaptation as development' operates at a different scale of theory and practice: that of the community.

42 *Adaptation, development, maladaptation*

> Community based adaptation (CBA) claims to identify, assist, and implement community-based development activities, research and policy in regions where adaptive capacity is as dependent on livelihood indicators as climatic changes.
>
> (Ayers & Dodman 2010, p. 164)

Currently, adaptation policies emphasise technocratic and managerial solutions to 'climate proofing' in development programmes as key to adaptation (Brooks et al. 2009; Head 2010; McMichael 2009). The technocratic paradigm deepens tendencies to override local knowledge and solutions. The argument here, and the critique, is that as the climate change crisis deepens, technocratic and managerial interventionist approaches are more suited to the challenge (Brooks et al. 2009), 'rather than support documented local initiatives based on adaptive practices, largely by women' (McMichael 2009, p. 252). The latter adaptation approach, in part, fits the emerging discourse on CBA (Ayers & Dodman 2010; Ayers & Forsyth 2009; Jones & Rahman 2007; Leopold & Mead 2009) that views 'adaptation as development'. That is, adaptation is local and place-based (Ayers & Dodman 2010). By contrast to the CBA approach, 'mainstreaming adaptation' views adaptation as 'adaptation plus development'. This approach views adaptation as external to development, that is something to be 'tacked on' to development rather than an integral part of development itself (Ayers & Dodman 2010). This latter approach, as the Tuvalu example cited above demonstrates, is at best piece-meal. Indeed, this embodies another pathway to maladaptation, insofar as it promotes path dependency. That is, this approach 'sets paths that limit the choices available to future generations' (Barnett & O'Neill 2010, p. 211).

(3) Are current approaches to adaptation leading to maladaptation?

Are these approaches or types of adaptation policies under the UNFCCC and development communities leading to maladaptation? Answering that question has necessarily involved excavating a layer of epistemological divergence between the two disciplines dominating climate change – the natural sciences and the social sciences (Huq et al. 2006). There is an urgent need to ensure that adaptation policies pursued today do not lead to tomorrow's maladaptation. This requires implementing adaptation policies that rely on the knowledge and practices of local peoples affected by climate change. Indeed, results from the Indonesian study present findings that support this claim.

Conclusion

In conclusion, I have presented adaptation policies, such as 'precautious adaptation' as well as other variations from the development community, that advocate building development around the environment as alternatives to the current market-driven approaches that leave an unsustainable development

Adaptation, development, maladaptation 43

paradigm intact. That is, where business-as-usual approaches side-line alternative ways of organising economies, such as the notion of degrowth and steady-state economics that directly challenge the status quo, and leave the underlying causes of climate change unaddressed.

Note

1 'Sir Nicholas Stern, author of the British Government's Stern Report on Climate Change, has meanwhile championed the initiative of his private firm, IDEACarbon, to set up a carbon credit ratings agency – which many observers are likely to see as subject to the same type of conflict of interest that earlier afflicted Moody's and other credit ratings agencies that depended for their income on the companies whose products they were rating' (Lohmann 2009, p. 184).

References

Ayers, J & Dodman, D 2010, 'Climate change adaptation and development I: the state of the debate', *Progress in Development Studies*, vol. 10, no. 2, pp. 161–168.

Ayers, J & Forsyth, T 2009, 'Community-based adaptation to climate change', *Environment: Science and Policy for Sustainable Development*, vol. 51, no. 4, pp. 22–31.

Barnett, J & O'Neill, S 2010, 'Maladaptation', *Global Environmental Change*, vol. 20, pp. 211–213.

Bello, W 2009, 'Carbon markets: a fatal illusion', in S Bohm & S Dabhi (eds), *Upsetting the offset: the political economy of carbon markets*, e-book, accessed 21 May 2017, http://ma yflybooks.org/wp-content/uploads/2010/07/9781906948078UpsettingtheOffset.pdf.

Bond, P 2011, *Patrick Bond speaking at OccupyCOP17*, 10 December, online video, accessed 12 December 2011, https://youtu.be/gX2Pn6b_YuQ.

Boyd, E, Hultman, N, Roberts, J, Corbera, E, Cole, J, Bozmoski, A, Ebeling, J, Tippman, R, Mann, P, Brown, K & Liverman, D 2009, 'Reforming the CDM for sustainable development: lessons learned and policy futures', *Environmental Science & Policy*, vol. 12, no. 7, pp. 820–831.

Bronen, R 2009, 'Forced migration of Alaskan indigenous communities due to climate change: creating a human rights response', in A Oliver-Smith & X Shen (eds), *Linking environmental change, migration and social vulnerability*, UNU Institute for Environment and Human Security (UNU-EHS), e-book, accessed 21 May 2017, www. munichre-foundation.org/dms/MRS/Documents/Source2009_OliverSmith_ShenEnviron mentalChange_Migration.pdf.

Brooks, N, Grist, N & Brown, K 2009, 'Development futures in the context of climate change: challenging the present and learning from the past', *Development Policy Review*, vol. 27, no. 6, pp. 741–765.

Brown, K 2011, 'Sustainable adaptation: an oxymoron?', *Climate and Development*, vol. 3, no. 1, pp. 21–31.

Bumpus, A & Liverman, D 2008, 'Accumulation by decarbonisation and the governance of carbon offsets', *Economic Geography*, vol. 84, no. 2, pp. 127–155.

Burton, I 1997, 'Vulnerability and adaptive response in the context of climate and climate change', *Climatic Change*, vol. 36, no. 1, pp. 185–196.

Burton, I 2009, 'Deconstructing adaptation … and reconstructing', in ELF Schipper & I Burton (eds), *The Earthscan reader on climate change*, Earthscan, London, p. 11.

44 Adaptation, development, maladaptation

Castree, N, Demeritt, D, Liverman, D & Rhoads, B (eds) 2009, *A companion to environmental geography*, John Wiley & Sons.

Collett, L 2009, *A fair-weather friend? Australia's relationship with a climate-changed Pacific*, Australia Institute, accessed 21 May 2017, www.tai.org.au/node/1499.

Elliott, L 2007, 'Improving the global environment: policies, principles and institutions', *Australian Journal of International Affairs*, vol. 61, no. 1, pp. 7–14.

Eriksen, S, Aldunce, P, Bahinipati, CS, Martins, RDA, Molefe, JI, Nhemachena, C, O'Brien, K, Olorunfemi, F, Park, J, Sygna, L & Ulsrud, K 2011, 'When not every response to climate change is a good one: identifying principles for sustainable adaptation', *Climate and Development*, vol. 3, no. 1, pp. 7–20.

Fraser, N 2009, *Scales of justice: reimagining political space in a globalising world*, Polity Press, Cambridge, UK.

Gender CC 2011, *Climate Finance*, Gender CC: Women for Climate Change, accessed 10 June 2017, http://gendercc.net/genderunfccc/topics/climate-finance.html.

Global Environment Facility (GEF) Secretariat 2009, *Focal area: climate change*, Global Environment Facility, accessed 14 June 2017, www.thegef.org/sites/default/files/p ublications/ClimateChange-FS-June2009_2.pdf.

Goldtooth, TBK 2009, 'The REDD train is going pretty fast and it's left us at the station', accessed 8 June 2017, www.redd-monitor.org/2009/01/14/interview-with-tom-bk-gold tooth/.

Gray, J 2007, *Straw dogs: thoughts on humans and other animals*, 1st American paperback edn, Farrar, Straus & Giroux, New York.

Hall, N 2016, *Displacement, development, and climate change: international organisations moving beyond their mandates*, Routledge.

Harvey, D 2003, *The new imperialism*, Oxford University Press Inc., New York, pp. 137–182.

Head, L 2010, 'Cultural ecology: adaptation-retrofitting a concept?', *Progress in Human Geography*, vol. 34, no. 2, pp. 234–242.

Hennessy, K, Fitzharris, B, Bates, BC, Harvey, N, Howden, SM, Hughes, L, Salinger, L & Warrick, R 2007, 'Australia and New Zealand', *Climate Change 2007: Impacts, Adaptation and Vulnerability. Contribution of Working Group II to the Fourth Assessment Report of the Intergovernmental Panel on Climate Change*, accessed 10 May 2017, www.ipcc.ch/pdf/assessment-report/ar4/wg2/ar4-wg2-chapter11.pdf.

Huq, S, Reid, H & Murray, L 2006, *Climate change and development links*, Gatekeeper Series 123, International Institute for Environment and Development, e-book, accessed 10 June 2017, http://lib.icimod.org/record/12247/files/4750.pdf.

Jones, R & Rahman, A 2007, 'Community-based adaptation', *Tiempo*, vol. 64, pp. 17–19.

Juhola, S, Glaas, E, Linner, B & Neset, T 2016, 'Redefining maladaptation', *Environmental Science & Policy*, vol. 55, January, pp. 135–140.

Kallis, G 2011, 'In defence of degrowth', *Ecological Economics*, vol. 70, no. 5, pp. 873–880.

Kates, R 2009, 'Cautionary tales: adaptation and the global poor', in ELF Schipper & I Burton (eds), *The Earthscan reader on climate change*, Earthscan, London.

Katz, C 2001, 'On the grounds of globalization: a topography for feminist political engagement', *Signs: Journal of Women in Culture and Society*, vol. 26, no. 4, pp. 1213–1234.

Lang, C 2009, 'Forests, carbon markets and hot air: why the carbon in forests should not be traded', in S Bohm & S Siddhartha (eds), *Upsetting the offset: the political economy of carbon markets*, e-book, accessed 21 May 2017, http://mayflybooks.org/ wp-content/uploads/2010/07/9781906948078UpsettingtheOffset.pdf.

Lang, C 2012, 'Down the rabbit hole: REDD in the CDM', *REDD-Monitor*, weblog post, 12 September, accessed 10 May 2017, www.redd-monitor.org/2012/09/12/down-the-rabbit-hole-redd-in-the-cdm/#more-12907.

Lang, C 2013, 'Clean Development mechanism: zombie projects, zero emission reductions and almost worthless carbon credits', *REDD-Monitor*, weblog post, 12 July, accessed 10 May 2017, www.redd-monitor.org/2013/07/12/clean-development-mechanism-7000-projects-zero-emissions-reductions-almost-worthless-carbon-credits-and-zombie-projects-increasing/.

Latouche, S 2010, 'Degrowth', *Journal of Cleaner Production*, vol. 18, no. 6, pp. 519–522.

Leopold, A & Mead, L 2009, 'Third international workshop on community-based adaptation to climate change', *Community Based Adaptation Climate Change Bulletin*, vol. 135, no. 2, accessed 10 June 2017, http://enb.iisd.org/download/pdf/sd/ymbvol135num2e.pdf.

Liverman, D 2009, 'Conventions of climate change: constructions of danger and the dispossession of the atmosphere', *Journal of Historical Geography*, vol. 35, no. 2, pp. 279–296.

Liverman, D 2011, *Amsterdam Global Climate Institute, No1-4.wmv*, uploaded 19 September 2011, online video, accessed 16 May 2017, www.youtube.com/watch?v=O5Nl0Gviz9s.

Lohmann, L 2008, 'Carbon trading, climate justice and the production of ignorance: ten examples', *Development*, vol. 51, no. 3, pp. 359–365.

Lohmann, L 2009, 'Regulation as corruption in the carbon offset markets', in S Bohm & S Siddhartha (eds), *Upsetting the offset: the political economy of carbon markets*, e-book, accessed 21 May 2017, http://mayflybooks.org/wp-content/uploads/2010/07/9781906948078UpsettingtheOffset.pdf.

Lovelock, J 2006, *The revenge of Gaia: why the earth is fighting back and how we can still save humanity*, Allen Lane, Camberwell.

Lovelock, J 2009, *The vanishing face of Gaia: a final warning*, Allen Lane, London.

Magnan, AK, Schipper, ELF, Burkett, M, Bharwani, S, Burton, I, Erikson, S, Gemenne, F, Schaar, J & Zievogel, G 2016, 'Addressing the risk of maladaptation to climate change', *Wiley Interdisciplinary Reviews: Climate Change*, vol. 7, September/October, pp. 646–665.

McMichael, P 2009, 'Contemporary contradictions of the global development project: geopolitics, global ecology and the "development climate"', *Third World Quarterly*, vol. 30, no. 1, pp. 247–262.

Noble, I, Huq, S, Anokhin, Y, Carmin, J, Goudou, D, Lansigan, F, Osman-Elasha, B & Villamizar, A 2014, 'Adaptation needs and options', in C Field, V Barros, D Dokken, K Mach, M Mastrandrea, T Bilir, M Chatterjee, K Ebi, Y Estrada, R Genova, B Girma, E Kissel, A Levy & S MacCracken (eds), *Climate Change 2014: impacts, adaptation, and vulnerability, part A: global and sectoral aspects, contribution of Working Group II to the Fifth Assessment Report of the Intergovernmental Panel on Climate Change*, Cambridge University Press, Cambridge, United Kingdom and New York, pp. 833–868.

Okereke, C 2007, *Global justice and neoliberal environmental governance: ethics, sustainable development and international co-operation*, 1st edn, Taylor and Francis, Florence.

Paulsson, E 2009, 'A review of the CDM literature: from fine-tuning to critical scrutiny?', *International Environmental Agreements: Politics, Law and Economics*, vol. 9, no. 1, pp. 63–80.

46 *Adaptation, development, maladaptation*

Pielke, R 2009, 'Rethinking the role of adaptation in climate policy', in ELF Schipper & I Burton (eds), *The Earthscan reader on climate change*, Earthscan, London, p. 345.

Prins, G & Rayner, S 2007, 'Time to ditch Kyoto', *Nature*, vol. 449, no. 7165, pp. 973–975.

Schalatek, L, Bird, N & Brown, J 2010, 'Where's the money? the status of climate finance post-Copenhagen', *Climate Finance Policy Brief*, no. 1, accessed 6 June 2017, www.odi.org/sites/odi.org.uk/files/odi-assets/publications-opinion-files/5844.pdf.

Scheraga, J & Grambsch, A 1998, 'Risks, opportunities, and adaptation to climate change', *Climate Research*, vol. 11, no. 1, pp. 85–95.

Schipper, ELF 2006, 'Conceptual history of adaptation in the UNFCCC process', *Review of European Community & International Environmental Law*, vol. 15, no. 1, pp. 82–92.

Schipper, ELF 2009, 'Conceptual history of adaptation in the UNFCCC process', in ELF Schipper & I Burton (eds), *The Earthscan reader on climate change*, Earthscan, London, pp. 359–376.

United Nations Framework Convention on Climate Change (UNFCCC) 1992, United Nations General Assembly, 20 January 1994, accessed 9 May 2017, www.refworld.org/docid/3b00f2770.html.

United Nations Framework Convention on Climate Change (UNFCCC) Adaptation Fund 2011, *Adaptation Fund*, United Nations Framework Convention on Climate Change, accessed 30 June 2011, http://unfccc.int/kyoto_protocol/mechanisms/clean_development_mechanism/items/2718.php.

United Nations Framework Convention on Climate Change (UNFCCC) CDM 2011, *Clean Development Mechanism (CDM)*, United Nations Framework on Climate Change, accessed 30 June 2011, http://unfccc.int/kyoto_protocol/mechanisms/clean_development_mechanism/items/2718.php.

Whiteside, K 2006, *Precautionary politics: principle and practice in confronting environmental risk*, MIT Press, Cambridge, Massachusetts.

Work, C, Rong, V, Song, D & Scheidel, A 2019, 'Maladaptation and development as usual? Investigating climate change mitigation and adaptation projects in Cambodia', *Climate Policy*, vol. 19, no. 51, pp. 547–562.

4 Mitigation and the Kyoto CDM
Manufacturing maladaptation

Introduction

Recognition of the existence of a clear link between climate change and displacement has seen environmental change placed firmly on the United Nations Framework Convention on Climate Change (UNFCCC) Conference of the Parties (COP) agenda as a specific category of displacement. Under the Cancun Adaptation Framework adopted at the UNFCCC COP 16 in Cancun 2010, the parties resolved to enhance understanding of climate change-induced displacement, planned relocation and migration. Further, this enhanced understanding would be done cooperatively and in a coordinated manner at appropriate international, national and regional levels (UNFCCC COP 16 AWG-LCA 2010, paragraph 14(f)). The link between maladaptation arising from mitigation policy, thus leading to potential migration and displacement, has not been adequately or theoretically conceptualised.

In the previous chapter on the one and many faces of adaptation, historically, climate change adaptation under the UNFCCC/Kyoto process had been the overlooked cousin of mitigation policy (Schipper 2006). This changed in recent years due to several factors alluded to by Pielke et al. (2007), namely: time-scale mismatch, that is, irrespective of decarbonisation of the global energy system, historical emissions dictate climate change is unavoidable; for reasons that have nothing to do with greenhouse gas emissions, vulnerability to climate-related impacts is caused by socioeconomic inequity, unsustainable patterns of development as well as rapid population growth; and, for those most at risk from climate change impacts and those least responsible for global warming, vulnerable developing countries are demanding the focus be on increasing resilience to climate events that will occur regardless of mitigation.

The following pages present a detailed analysis of how climate change mitigation policy under the UNFCCC Kyoto Protocol's Clean Development Mechanism (CDM) and its successor mechanism under Article 6.4 of the Paris Agreement, is leading to maladaptation. What emerges from this critique is that in its current form as an international crediting mechanism, the CDM is operating more like a mechanism for manufacturing maladaptation than mitigation.

48 *Mitigation and the Kyoto CDM*

Introduction to the Clean Development Mechanism

The CDM is one of three 'flexibility mechanisms', in addition to Joint Implementation and Emissions Trading, under the Kyoto Protocol that allow wealthy industrialised countries (Annex 1 Parties) with greenhouse gas (GHG) reduction targets to carry out emission mitigation projects in developing countries (non-Annex 1 Parties) at the lowest possible cost (Boyd et al. 2008). Thus, the CDM is a carbon offset mechanism. Bumpus and Liverman (2008) state that while there is no consensus or agreement on a general definition of a carbon offset, they define one as 'occurring when one actor ... invests in a project elsewhere that results in a reduction of greenhouse gas emissions that would not have occurred in the absence of the project' (Bumpus & Liverman 2008, p. 135).

According to its advocates, the rationale for the CDM is predicated on a neoliberal economic view that the marginal cost for GHG emission reductions in developing countries is generally less than in industrialised countries (Solomon & Heiman 2010). That is, it is cheaper and easier than domestic reductions undertaken in Annex 1 countries and the projects also contribute to sustainable development (Bumpus & Liverman 2008). Emissions Trading and Joint Implementation, the latter of which is also an offset mechanism, are currently operational in developed countries, while the CDM allows developing countries, not subject to the Kyoto Protocol, and developed countries to jointly reduce emissions (Bohm & Dabhi 2009). Collectively, these mechanisms are part of what forms the global carbon market.

This analysis does not focus on the voluntary carbon market which falls outside of the compliance market associated with the UNFCCC/Kyoto regime and thus emission reductions generated in this voluntary market do not count under the Kyoto Protocol's binding target emissions for signatory Annex 1 countries (Bumpus & Liverman 2008). Rather, the analysis looks at how the CDM leads to maladaptation, thus presenting as a different driver, like climate change itself, for the potential migration and displacement of people.

The CDM is established under Article 12 of the Kyoto Protocol where Article 12.2 defines its purpose as follows:

> The purpose of the clean development mechanism shall be to assist Parties not included in Annex 1 in achieving sustainable development and in contributing to the ultimate objective of the Convention, and to assist Parties included in Annex 1 in achieving compliance with their quantified emission limitation and reduction commitments under Article 3.
>
> (UNFCCC Kyoto Protocol 1998)

There has been a substantial body of work that has focused on whether the CDM has been able to fulfil the dual objective required by the Kyoto Protocol. Early evidence in the lead up to and during the first commitment period of the Kyoto Protocol (2008–2012), found that many of the CDM projects lack

Mitigation and the Kyoto CDM   49

sustainable development benefits while the mechanism has been effective in achieving its cost-efficient emission reduction objective (Boyd et al. 2009; Bozmoski et al. 2008; Bumpus & Cole 2010; Najam et al. 2003; Olsen 2007). Sutter & Parreno's (2007) early analysis of the CDM found that while 72 per cent of the 16 officially registered CDM projects under study, 'expected Certified Emission Reductions (CERs) are likely to represent real and measurable emission reductions, less than 1% are likely to contribute significantly to sustainable development in the host country' (Sutter & Parreno 2007, p. 75). With the passage of time, real and measurable emission reductions increasingly began to evaporate, calling into question the CDM's environmental integrity as well.

The CDM was the subject of a more recent, comprehensive study entitled *How additional is the Clean Development Mechanism?*, conducted by the Öko Institute, Institute of Applied Ecology, in partnership with INFRAS and the Stockholm Institute of Environment. Their results, based on a detailed analysis of 300 project types randomly drawn from the CDM project portfolio in 2014 (CDM pipeline), found '85% of the projects covered in this analysis and 73% of potential 2013–2020 Certified Emission Reductions (CERs) supply have a low likelihood that emission reductions are additional and are not overestimated' (Cames et al. 2016, p. 11). Unlike the earlier studies on the CDM undertaken in the first commitment period which focused on the CDM's sustainable development objective, their study did not include this in their analysis.

In spite of equal priority being given to both objectives in the Kyoto text (Bozmoski et al. 2008), many authors have argued there is an inherent bias in the CDM that privileges low-cost emission reductions, or cost-effectiveness, at the expense of sustainable development. That is because 'sustainable development has no monetary value in the [CDM] mechanism' (Paulsson 2009, p. 70). Additionally, sustainable development has no operational definition (Boyd et al. 2009). The monetary value in the CDM derives from making greenhouse gas reductions (Andrew et al. 2010) into a commodity – in this case a CER. If, for example, an Annex 1 country, inclusive of 'private and/or public entities' (UNFCCC Kyoto Protocol 1998, Article 12.9) is unable or unwilling to reduce its GHG emissions efficiently, that is because it would cost it too much, then it could simply buy CERs from a country or company in the South (non-Annex 1 Party) which would have a comparative cost advantage in implementing GHG emission cuts (Bohm & Dabhi 2009).

The debate on how the CDM can be reformed to deliver on its sustainable development objective (Boyd et al. 2009; Bumpus & Cole 2010) and its twin objective of environmental integrity, notwithstanding 'fundamental and far-ranging changes to the CDM' being needed (Cames et al. 2016), is well established. What has been less discussed since its inception is whether the CDM is potentially leading to maladaptation. If this is the case, then not only does this undermine the CDM's stated sustainable development objective which 'is needed because it can provide the conditions in which climate policies can be implemented' (Najam et al. 2003, p. 228), but it also brings into question the integrity of GHG emission reductions under this offset mechanism. In this case,

50 *Mitigation and the Kyoto CDM*

the CDM would seem to be beyond reform and no amount of 'fine-tuning' of this mechanism (Paulsson 2009) will remedy it.

In a critical review of the CDM literature, Paulsson (2009) found that only a 'handful of articles' were written by authors eschewing the policy-oriented approach that has dominated the CDM literature by taking a step back and looking at the CDM from a broader, more critical and theoretical, perspective. Using Cox's (1981) distinction between 'problem-solving' and 'critical', Paulsson (2009) found that most of the CDM literature reviewed could be classified as the former. While the usefulness of problem-solving research may be appropriate in the process of day-to-day policy-making, this approach works within the current neoliberal paradigm, and thus the prevailing system is never brought into question (Paulsson 2009). Those critical authors calling the prevailing neoliberal system into question were further distinguished by two types: those 'very critical voices [who] question the whole idea of a CDM mechanism in the climate regime ... and [those who] look at the CDM through a theoretical lens' (Paulsson 2009, p. 76).

The critical theoretical approach used in this chapter is a synthesis of both types, employed to conceptually advance the argument that it is not further reform of the CDM that is needed but a radical transformation (Cames et al. 2016) to avoid maladaptation. This has important implications for finalising rules for the new mechanism under Article 6.4 of the Paris Agreement at the COP 26 to be held in Glasgow in November 2021 (previously November 2020, but now rescheduled due to the ongoing COVID-19 pandemic). It calls for the Conference of the Parties serving as the meeting of the Parties to the Paris Agreement (CMA) to put 'the risk of maladaptation at the top of the planning agenda ... starting with an intention to avoid mistakes and not lock-in detrimental effects' (Magnan et al. 2016, p. 646), as its predecessor has demonstrated.

In its current form, the CDM will end when the second commitment period of the Kyoto Protocol concludes in 2020 (Cames et al. 2016). The new mechanism established under Article 6.4 of the Paris Agreement to mitigate GHG emissions is near to final draft. At the COP 25 UN climate talks in Madrid, December 2019, the main aim was to finalise the Paris Agreement 'rulebook', by setting rules of carbon markets and other forms of international cooperation under 'Article 6' of the Agreement. 'These "Article 6" rules are the last piece of the Paris regime to be resolved, after the rest of its "rulebook" was agreed to in late 2018 [at COP 24 in Katowice, Poland]' (Evans & Gabbatiss 2019a). The near-final draft of the Article 6.4 rules at COP 25, 'would have allowed the use of CERs within the Paris regime, but with a "vintage" restriction, meaning only credits created after a certain date would be allowed' (Evans & Gabbatiss 2019b). That date, according to the draft text of Article 6.4, paragraph 75(a), was to be determined later, 'meaning that unless agreement was reached then no CERs would be allowed' (Keohane 2019, cited in Evans & Gabbatiss 2019b). A significant, and ultimately unresolved, contention for the CMA in Madrid in 2019, was 'how to deal with billions of Kyoto-era carbon offset "units", potentially amounting to more than five billion tonnes of CO_2 equivalent' (Evans & Gabbatiss 2019b). China, India and Brazil hold the 'lion's share' of these units, mostly generated under the CDM as CERs. 'At COP 25, these countries pushed for CERs to be eligible under Article 6.4, arguing that private companies had invested in good faith and should not have their assets rendered worthless' (Evans & Gabbatiss 2019b).

If the CMA, in finalising the new mechanism under Article 6.4, is to learn lessons from the well documented inherent flaws of the CDM, it should heed the lesson that 'after well over a decade of gathering considerable experience, the enduing limitations of GHG crediting mechanisms [CDM] are apparent ... elusiveness of additionality [is but one of a number of significant limitations]' (Cames et al. 2016, p. 17). Therefore, it is crucial that the CMA at the COP 26 in Glasgow in November 2021 prioritise the risk of maladaptation, by not locking in negative effects, thus avoiding path dependency, the final pathway through which maladaptation arises under the Barnett and O'Neill typology. Cames et al.'s (2016) study on the CDM, published in 2016, included important insights to improve the CDM up to 2020, suggesting the approach 'could also be applied more generally to assess the environmental integrity of other compliance offset mechanisms, as well as to avoid flaws in the design of new mechanisms being used or established for compliance' (Cames et al. 2016, p. 10). The authors 'recommend focussing climate mitigation efforts on forms of carbon pricing that do not rely extensively on credits' (Cames et al. 2016, p. 11).

Endless attempts to overcome shortcomings identified in the CDM have focused on a problem-solving approach (Paulsson 2009) to remedy fundamental flaws that have persisted throughout its life, manifest in the notion that if we try a bit harder, with a little more fine-tuning here and there, more regulation, better governance, overcome the problem of finding an operational definition for sustainable development in the CDM, improve monitoring and enforcement, better verification of additionality (not business as usual) of CDM projects – it will work. The 'baby' will be 'the win-win instrument', 'the front-runner of the Kyoto Regime' and 'a bridge between North and South' that 'the Kyoto surprise' promised to be when it was born in 1997 (Grubb 1999, Matsuo 2003, cited in Olsen 2007, p. 61). The question of whether an offset crediting mechanism, like the CDM in its current form, can safeguard climate stability, or the Earth's capacity to regulate climate, by making it a commodity, an asset (Lohmann 2009a, 2009b), has always been a dangerous and misguided neoliberal chimera. In establishing a series of market-based mechanisms under the rubric of the 'carbon markets' to combat global warming, the Kyoto Protocol has been framed in the shadows of neoliberalism (Dabhi 2009).

According to McCarthy and Prudham (2004), in spite of the ubiquity of the term neoliberalism, defining it is no straightforward task (McCarthy & Prudham 2004). As a mode of discourse, neoliberalism has become hegemonic (Harvey 2007), and

52 *Mitigation and the Kyoto CDM*

> is made most evident by the ways in which profoundly political and ideological projects have successfully masqueraded as a set of objective, natural, and technocratic truisms [market-based climate change solutions]. Political resistance gives the lie to such disguises, exposing the political negotiations and myriad contradictions, tensions, and failures of neoliberalisations.
>
> (McCarthy & Prudham 2004, p. 276)

If the Kyoto Protocol was formed in the shadows of neoliberalism (Dabhi 2009), then the aim of critiquing the Kyoto CDM involves defining and deconstructing neoliberalism's central tenets in their relationship to accelerating the production of environmental risks (McCarthy & Prudham 2004) and the erosion of the preconditions for sustainable development, thus leading to maladaptation (Juhola et al. 2016). This critical analysis of the CDM employs the conceptual and theoretical resources of a wide inter/trans-disciplinary field which spans the sciences and humanities, including critical geography (McAfee 1999; McCarthy & Prudham 2004; Castree 2006; Bumpus & Liverman 2008; Liverman 2009; Silvey 2009), critical accounting (Andrew et al. 2010), and ecological economics (Brown 2013; Daly 1995, 2005, 2008; Daly & Townsend 1993; Kosoy & Corbera 2010; Muradian et al. 2010). What these various disciplines hold in common is a deep scepticism that the dominant ideology, discourse and practice of neoliberal environmentalism in general, and the market solutions established under the Kyoto Protocol in particular, can 'solve' climate change. 'Climate change presents a challenge that will never be "solved" … we can [only] do better or worse in our managing of it' (Prins et al. 2010, p. 36). We have to adapt to climate change whether we want it or not (Adger et al. 2006).

Opponents of the current CDM have labelled it variously a 'dangerous', 'fatal illusion', 'big delaying tactic', 'false solution' or 'dangerous diversion' (Bello 2009; Goldtooth 2010; Smith 2008). As this analysis lays bare, the CDM is the policy mechanism facilitating maladaptation. Superficially, maladaptation may appear to be an unintended consequence of the CDM, but what the analysis brings to the surface is how this offset mechanism ensures that the dominant ideology of neoliberalism serves to obscure and thus keep intact the dominant market-based logic that underscores climate change solutions. Just as 'short-term adaptation can lead to long-term maladaptation' (Brooks et al. 2009, p. 741), similarly, today's mitigation policy can lead to tomorrow's maladaptation. Thus, the potential migration and displacement of peoples due to maladaptation is a spatio-temporal manifestation of neoliberal market environmentalism, whose dimensions remain unknown but who owe their existence to maladaptive climate change solutions.

What neoliberalism, variously described as ideology, material practice and discourse (McCarthy & Prudham 2004), fails to recognise within its core tenets of individual liberty, unencumbered markets, private property rights and free trade (Harvey 2005), is that '[l]andscapes of carbon production are much more complicated than in linear carbon trading ... individual [CDM] projects are deeply conditioned by people and places, in which they are cited' (Stewart 2010, p. 38). The free-market orientation of contemporary neoliberal solutions to climate change (Andrew et al. 2010), what Solomon and Heiman (2010) describe as a 'modern day GHG indulgence system', thus downplays neoliberalism as an 'extra-local project' (McCarthy & Prudham 2004).

In other words, neoliberalism as a totalising, ideological hegemonic mode of discourse tends to obscure the material practical effects at the local level. For example, the CDM or its Paris successor may serve as mechanisms for global mitigation of GHG emissions. But neoliberalism as a material practice disproportionately burdens the most vulnerable at the local level in which CDM projects are located. Too often the voices of indigenous peoples, local communities and civil society actors have been marginalised in the set-up and implementation of specific projects, and the evidence presented earlier about the CDM's sustainable development objective not being met at the expense of cost-effective abatement serves, in part, to confirm this.

Does the CDM lead to maladaptation?

One of the key aims here is to assess whether the CDM is leading to maladaptation, and thus presenting as a different driver for displacement of peoples. Barnett and O'Neill's (2010) typology of maladaptation in the context of climate change adaptation is used to frame this critique. That is, maladaptation is usually associated with adaptation, and their typology, one of the earliest attempts to conceptualise maladaptation, systematically, has since been followed by others (see Juhola et al. 2016; Magnan et al. 2016). All three conceptualisations of maladaptation focus on adaptation, which describes the ways in which adaptation actions or decisions to climate change can lead to the unintended consequences of maladaptation.

Six years after Barnett and O'Neill's typology, Juhola et al. (2016) expanded the definition to include 'intentional' actions as well. That is, 'maladaptation could be defined as a result of an intentional adaptation policy or measure directly increasing vulnerability for the targeted and/or external actor(s), and/or eroding preconditions for sustainable development by indirectly increasing society's vulnerability' (Juhola et al. 2016, p. 139). In this analysis, the theoretical and practical purchase of the maladaptation typology is broadened, to include mitigation policy as others have recently done (Work et al. 2019), on the basis that mitigation shares with adaptation 'a broad array of institutional practices, discourse and policies, and includes any and all activities created in the context of managing climate change (mitigation projects, for example)' (Work et al. 2019).

Additionally, justifications for extending maladaptation to include mitigation policy are found. Firstly, in the Intergovernmental Panel on Climate

54 Mitigation and the Kyoto CDM

Change (IPCC) assertion that in addition to adaptation as one of two human responses to climate change, mitigation constitutes the other. Thus 'mitigation and adaptation must go hand in hand' (Prins & Rayner 2007, p. 975). Secondly, and in connection with the first, mitigation policy and adaptation policy have a direct link with the CDM. That is, the CDM is the main source of funding for the Kyoto Adaptation Fund (UNFCCC Adaptation Fund 2011; UNFCCC CDM 2011) and remained the only trading mechanism 'taxed' for the adaptation funds (Boyd et al. 2009, p. 829) during the first commitment period. At COP 18 in Doha in 2012, this was extended to include the other two market mechanisms under the Kyoto Protocol, that is, international emissions trading and Joint Implementation during the second commitment period. The Adaptation Fund was established at COP 7 in 2001 to fund adaptation programmes and projects in developing countries that are Party to the Kyoto Protocol. The fund is financed with a share of the proceeds from CDM project activities that amount to 2 per cent of CERs issued for a CDM project activity (Camargo 2009).

> The CDM allows emission-reduction projects in developing countries to earn certified emission reduction (CER) credits, each equivalent to one tonne of CO_2. These CERs can be traded and sold, and used by industrialised countries to meet a part of their emission reduction targets under the Kyoto Protocol.
>
> (UNFCCC CDM 2011)

A third and final justification for extending the maladaptation typology to mitigation policy relates to an earlier justification discussed. That is, avoiding maladaptation embodies the principle of precaution or the precautionary principle (PP), which is embedded in Article 3.3 of the UNFCCC. The UNFCCC does not define maladaptation (Barnett & O'Neill 2010). However, in theory, the PP states that lack of full scientific certainty should not be a reason to postpone cost-effective measures to address climate change (Elliott 2007). It would appear then, the PP has been incorporated into one of neoliberal environmentalism's central articles of faith – cost-effectiveness or efficiency, to deal with climate change at the 'lowest cost possible' (UNFCCC, Article 3.3).

The current research suggests, however, the CDM still has fundamental flaws in terms of overall environmental integrity (Cames et al. 2016), thus defeating the 'ultimate objective' of the Climate Convention, which is the 'stabilisation of greenhouse gas concentrations in the atmosphere at a level that would prevent dangerous anthropogenic interference with the climate system' (UNFCCC 1992, Article 2). The lowest cost possible, in the long run, has resulted in the dual objectives of the CDM being plagued with issues of enduring integrity and legitimacy. Perspectives from the South warned that Kyoto was becoming more of a mechanism for managing global carbon trade than the more fundamental role of reducing emissions for atmospheric carbon stabilisation which could be neglected or delayed (Najam et al. 2003).

With these justifications for extending maladaptation to mitigation policy in place, Barnett and O'Neill's (2010) typology is used as a conceptual intervention to analyse how maladaptation arises from the Kyoto Protocol's CDM, and is seemingly embedded in each of the following five pathways. That is, relative to alternatives, maladaptation arises from actions that: 'increase emissions of greenhouse gases; disproportionately burden the most vulnerable; have high opportunity costs; reduce incentives to adapt and, set paths that limit the choices available to future generations [path dependency]' (Barnett & O'Neill 2010, p. 211).

(1) How the Kyoto CDM increases emissions of greenhouse gases

As the first pathway through which maladaptation arises – that is, an action, decision or policy that leads to an increase in GHG emissions. Early research in the first commitment period of the Kyoto Protocol (2008–2012) found that the CDM was failing to lead to additional emission reductions being met and that it was leading to a net increase in GHG emissions (Liverman 2009). The temporal dimension of mitigation policy is important here because, like adaptation actions, outcomes span across timescales (Juhola et al. 2015).

Almost a quarter of a century has passed since the European community and 37 industrialised countries (Annex 1 Parties under the UNFCCC) agreed in 1997 to reduce greenhouse gas emissions to an average of 5 per cent (5.2 per cent) against 1990 levels over the four-year period 2008–2012.[1] In addition to national efforts to achieve binding targets set by the Kyoto Protocol, the treaty offered Annex 1 Parties, known as Annex B countries under the Protocol, 'an additional means of meeting their targets by way of three market-based [flexibility] mechanisms' (UNFCCC Kyoto Protocol 1998), of which the CDM is one. Thus, a market-based mitigation solution to climate change was locked into international environmental law through the Kyoto Protocol and was perceived as a triumph for the neoliberal project of market environmentalism (Liverman 2009) – what others refer to as 'free-market' environmentalism (McCarthy & Prudham 2004).

During this time, there has been increasing evidence to suggest that carbon markets have assisted in legitimising real growth in carbon emissions; if some Annex B countries that have binding Kyoto targets to reduce GHG emissions appear to be meeting these, then, it is claimed, that is because this involves a lot of 'creative accounting' (Bohm & Dabhi 2009). The concept of 'additionality' is at the heart of the creative accounting claim made here, and it is this central condition introduced into the CDM, alongside sustainable development, that no CDM project would be allowed to progress without (Bohm & Dabhi 2009).

Article 12.5 of the Kyoto Protocol outlines what constitutes additionality over two key clauses. That is,

56 *Mitigation and the Kyoto CDM*

[e]mission reductions resulting from each [CDM] project activity shall be ...

b Real, measurable, and long-term benefits related to the mitigation of climate change; and

c Reductions in emissions that are additional to any that would occur in the absence of the certified project activity.

(UNFCCC Kyoto Protocol 2009)

The reason the concept of additionality was introduced by the UNFCCC was to eliminate the prospect of spurious CER credits being sold by developing countries to developed countries. That is because, unlike developed countries (Annex B Parties under the Kyoto Protocol), which have binding targets for reducing GHG emissions, developing countries that are signatories are not subject to Kyoto (Wysham 2008; Solomon & Heiman 2010). Therefore, trading under the CDM is problematic from the point of view that trading occurs between industrialised countries that have capped emissions and those developing countries that do not have caps on their emissions (Wysham 2008).

In order to avoid bogus emissions credits being sold by developing countries to Annex B countries, the UNFCCC decided that carbon credits would be issued under the CDM only if they were 'additional' – or 'not business as usual'. This concept of additionality, which requires the proof of a counter-factual, has been all but impossible to verify.

(Wysham 2008)

Figure 4.1 demonstrates the concept of additionality. That is, additionality distinguishes reductions in emissions generated by the CDM project from the business-as-usual baseline emissions without the project (Michaelowa 2005, cited in Bumpus & Liverman 2008). Additionality is thus represented in the figure as 'emissions reductions'.

Describing counterfactual (non-existing) futures scenarios by CDM project developers can lead to two types of additionality not being met – environmental additionality and financial additionality. For example, in the case of environmental additionality not being met, this occurs when baselines are exaggerated and claimed reductions in emissions have not actually occurred. With respect to no financial additionality being met by the proposed project, this occurs when carbon credits are generated by projects that would have happened anyway (Bumpus & Liverman 2008). These projects should therefore not qualify as CDM because they do not create additional emissions reductions.

In fact, they exacerbate the problem of climate change by allowing companies in the developed world, as the purchasers of CDM credits, to exceed their emission limits without genuinely offsetting elsewhere (Pearce 2008). In other words, when projects lack environmental additionality this allows for an

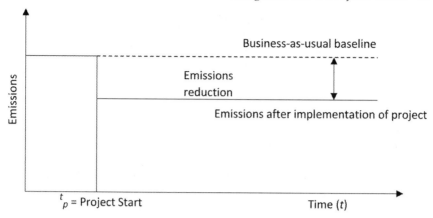

Figure 4.1 Simplified principle of the baseline (adapted from Michaelowa 2005). The emissions reduction gives the calculation for the amount of CO_2e (carbon dioxide equivalent) reduced as a result of the implementation of the project
Source: (Bumpus & Liverman 2008, p. 135) (used with permission).

increase in GHG emissions occurring somewhere else. This in turn leads to an overall net increase in emissions (Smith 2008, 2009). Offsetting, by definition, is designed to produce a 'zero-sum' outcome, not reducing emissions, but simply moving emissions where reductions are made (Gerebizza 2011). It appears, at least on the first pathway through which maladaptation arises, the CDM offset has led to an increase in emissions of GHG emissions through additionality not being met.

According to Pearce (2008), evidence in support of spurious emission-reduction CDM projects is 'alarming'. In 2007, an expert adviser to the CDM executive board said one-third of Indian CDM projects lacked additionality or were simply business-as-usual (Smith 2008, 2009; Lohmann 2009a, 2009b). David Victor from Stanford University claims emission reductions have not resulted from over two-thirds of CDM projects in developing countries that have been issued with credits (Wysham 2008). In another study by International Rivers, an international non-governmental organisation (NGO) that campaigns against hydroelectric dams, the vast majority of the many hundreds of hydro dams under construction in China issued with CERs were either under construction or completed before the application of carbon credits were made (Pearce 2008; Smith 2008, 2009). In other words, the projects were going to be developed anyway, regardless of carbon finance (Smith 2008, 2009).

According to Buiter (2007), the fact that offset accounting requires

> the impossible verification of how much carbon dioxide equivalent would have been emitted in some counterfactual alternative universe ... makes one shout out: impossible! Fraud! Bribery! Corruption! Wasteful diversion of resources into pointless attempts at verification!
> (Buiter 2007, cited in Lohmann 2009a, p. 182)

58 *Mitigation and the Kyoto CDM*

Counterfactual futures scenarios, alternative universes can be likened to what scientists call impossibility statements, the very foundation of science (Daly & Townsend 1993). For example, it is impossible to create or destroy matter-energy, travel faster than the speed of light or build a perpetual motion machine. In order to avoid wasting resources that will inevitably fail, they claim impossibility theorems should be heeded and respected by economists (Daly & Townsend 1993). Thus, the logic of impossibility theorems when applied to the CDM counterfactual construct is more than a waste of resources, it is part of a market solution to mitigating GHG emissions that is fundamentally and demonstrably flawed, that is in the meantime generating an invisible carbon bubble of alleged emission reductions that have never happened (Smith 2008, 2009).

Proponents argue, however, that to overcome the issue of spurious claims of emissions reductions being made under the UNFCCC Kyoto Protocol, the CDM has created many approved methodologies to determine baselines and an 'additionality tool' to assist project developers in overcoming the problem of creating projects that 'would not have happened anyway'. Theoretically, at least, these tools and methodologies assist project developers in being able to produce genuine additional emission reductions and thus generate credits that are credible (Bumpus & Liverman 2008). Approved methodologies under the CDM for large-scale projects, small-scale CDM projects, and Afforestation and Reforestation projects now cumulatively number 339 newly submitted approved methodologies as of March 2011 (UNEP-Risoe Centre CDM Pipeline 2011).

The idea of additionality being included in the CDM to begin with was to avoid spurious credits being claimed for emission reductions by developing countries that are Party to the Kyoto Protocol but do not have binding targets under the first compliance period from 2008 to 2012. Based on the foregoing, it would appear that in birthing the additionality condition, further reforms in the form of an 'additionality tool' and multiple methodologies have now been needed to eliminate other types of spurious claims being made to additionality itself, namely to projects that lack financial additionality and those that lack environmental additionality. This brings us back to the two broad types of research Paulsson (2009) distinguished in the CDM literature. That is, 'problem-solving' and 'critical', where the latter calls the prevailing system into question. However, problem-solving dominates much of the research. That is, where endless attempts to 'fine-tune' the CDM are made, and remade, in an attempt to justify the verifiability of additionality – a task that is proving difficult in the construct of a hypothetical world without the project.

There is, however, a more fundamental problem with offsets. That is, awarding CDM offset projects CERs for the difference between baseline and actual emissions will always, and only ever, be marginal. 'Ultimately, it is the [non-fabricated] baseline emissions path that must be altered if the problem of global warming is to be resolved' (Wara 2007, p. 596). Reinforcing this point, critical geographer Diana Liverman (2009) has estimated, based on several

Mitigation and the Kyoto CDM 59

modelling studies, that the flexibility mechanisms under the Kyoto Protocol 'will leave 450 million metric tonnes more carbon in the atmosphere by 2012 than if emission reductions were made domestically' (Liverman 2009, p. 279). By way of comparison, she adds, this is the equivalent effect of US non-participation in Kyoto. According to Carbon Market Watch, 'in Europe alone, the use of junk CDM credits increase emissions by about 750 million tonnes of CO2e' (Dufrasne 2018).

Alternative policy instruments, other than carbon offset mechanisms such as the CDM to address the baseline emissions path, are advanced by many critics of the dominant neoliberal approach driving emissions trading. These alternative approaches fall broadly under the banner of 'command and control' mechanisms whose proponents argue are the most effective way to reduce emissions and thus meet targets under the Kyoto Protocol (Bumpus & Liverman 2008; Solomon & Heiman 2010; Wysham 2008). Still others eschew altogether the Kyoto approach's focus on decarbonisation as the sole encompassing goal (Prins et al. 2010). What these alternative approaches and policy instruments offer is a more direct and transparent way in which emission reductions can be accounted for, without the added weight of a complicated and complex array of monitoring and verification tools and multiple methodologies that the CDM in its current form relies on.

For example, an upstream carbon tax, which is designed to force change in energy procurement and use (Solomon & Heiman 2010); a carbon tax that would influence taxpayer behaviour and orient economic and social activity towards carbon mitigation and/or reduction (Andrew et al. 2010); a low hypothecated (dedicated) 'inefficient' tax, in strictly theoretical terms, that does not seek to change behaviour as the previous 'carbon tax' does, but is more modest and specific in its strategic aim. That is, it would switch the political priority of governments' 'current preoccupation with emissions targets under the "Kyoto" regime to credible long-term global commitments and methods to invest in energy innovation' (Prins et al. 2010, p. 34).

These policy instruments, which do not feature crediting mechanisms, are alternatives to the existing climate mitigation policy under Kyoto that have been facilitated through the commodification of carbon. The alternatives rely on the intervention of the state, something theoretically antithetical to a neoliberal project where the alleged self-regulating market establishes the price for carbon emissions. However, the proponents of the Kyoto market-based mechanisms ignore the obvious fact that the political and ideological rejection of state interference is accompanied by the deeply contradictory endorsement of the establishment of property rights that accompany the generation of emission credits (CERs), and their commodification that is defended and created by the state to begin with (McCarthy & Prudham 2004).

Alternative policy instruments to the CDM are discussed in further detail under the final pathway through which maladaptation arises. Namely, actions that 'set paths that limit the choices available to future generations [path dependency]' (Barnett & O'Neill 2010, p. 211). As we have seen, embedded in

60 *Mitigation and the Kyoto CDM*

this first pathway are the seeds of path dependency, rooted in the ideological foreclosure of alternative command and control mechanisms under the Kyoto Protocol in 1997 that has led to an increase in greenhouse gas emissions under the CDM a quarter of a century later. To avoid the risk of further manufacturing maladaptation in the new mitigation mechanism under Article 6.4 of the Paris Agreement, the Parties must heed the lessons of its predecessor, that have been gained from well over a decade of gathering considerable experience of the inherent failings and limitations of the CDM, and crediting mechanisms in general (Cames et al. 2016).

(2) How the CDM disproportionately burdens the most vulnerable

'Adaptation [mitigation] actions are maladaptive if, in meeting the needs of one sector [region] or group, they increase the vulnerability of those most at risk, such as minority groups' (Barnett & O'Neill 2010, p. 212). Under this pathway, the focus is to demonstrate how the CDM on an international regional scale is skewed in favour of the more developed of the developing countries. Consequently, Agarwal (1999, cited in Najam et al. 2003) claimed the poorest countries would be unable to benefit from the proposed solution in the Kyoto Protocol, that is, participation in carbon trading through the CDM, because they were not likely to attract private-sector funding in any case. The following charts demonstrate how the CDM has created a disproportionate burden on developing countries in favouring the more developed of the developing countries in the number of projects that are in the CDM pipeline. Banuri and Gupta (2000, cited in Najam et al. 2003) presciently warned that this would be the path the CDM would follow. That is, in 2000, they foresaw the likelihood of the CDM following the path of foreign direct investment where 'the much-trumpeted benefits [CERs] will accrue to a handful of the larger developing countries, leaving the bulk of the south on the side-lines of the global carbon market' (Banuri & Gupta 2000, cited in Najam et al. 2003, p. 225). Chile has usually used CDM projects to attract foreign investments, while little effort has been made to ensure sustainable development outcomes (Rindefall, Lund & Stripple 2011, cited in Mathur et al. 2014). Two decades later, China, Brazil and India, 'a handful of countries', hold the 'lion's share' of CDM CERs created from projects in their countries (Evans & Gabbatiss 2019b).

The following graphs illustrate how a 'handful of larger developing countries' benefitted from the CDM as predicted, in the first commitment period (2008–2012) and second commitment period (2013–2020). The 'intra regional country analysis presents an even more skewed geographical pattern of CDM projects and the volume of CERs awarded in the UNEP Risoe Pipeline' (last updated 1 March 2011, www.cdmpipeline.org/cdm-projects-re gion.htm#2).

In other words, Figure 4.2 indicates India and China host most of the projects in Asia, but most CERs are expected from China (Figure 4.4).

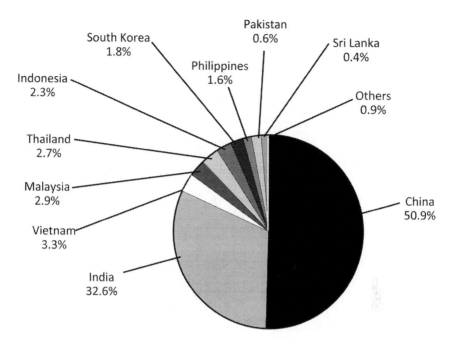

Figure 4.2 Number of CDM projects in Asia by country as at 1 March 2011

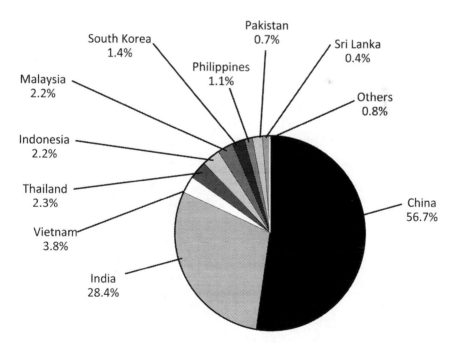

Figure 4.3 Number of CDM projects in Asia by country as at 1 August 2020

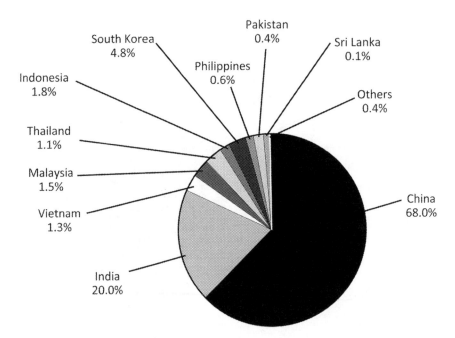

Figure 4.4 Volume of CERs until 2012 in Asia by country as at 1 March 2011

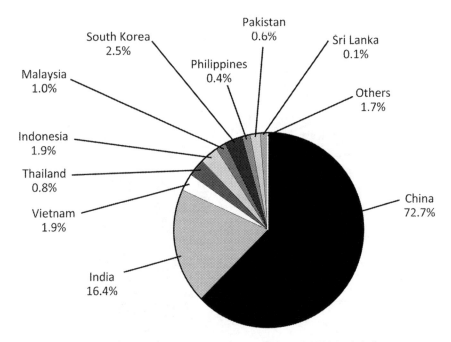

Figure 4.5 Volume of CERs for Commitment Period 2 until 2020 in Asia by country as at 1 August 2020

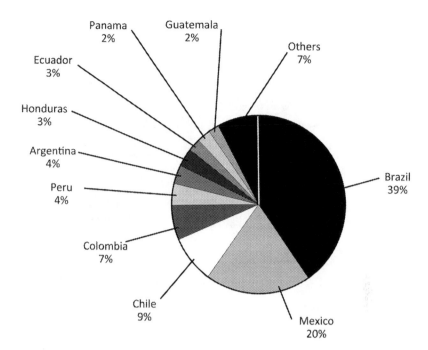

Figure 4.6 Number of CDM projects in Latin America by country as at 1 March 2011

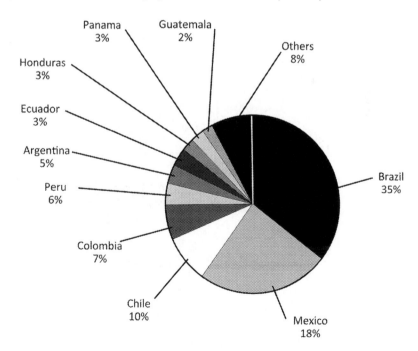

Figure 4.7 Number of CDM projects in Latin America by country as at 1 August 2020

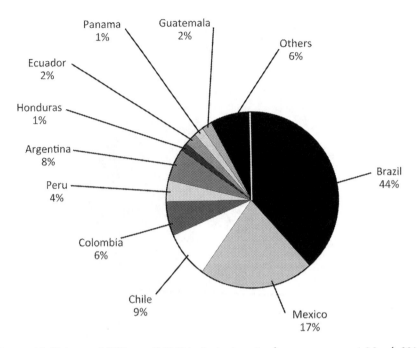

Figure 4.8 Volume of CERs until 2012 in Latin America by country as at 1 March 2011

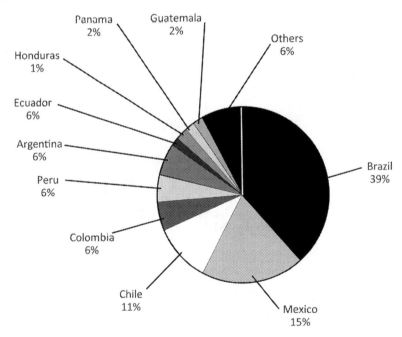

Figure 4.9 Volume of CERs for Commitment Period 2 until 2020 in Latin America by country as at 1 August 2020

Mitigation and the Kyoto CDM 65

(3) The high opportunity costs of the CDM

Relative to alternatives, if social, economic and environmental costs are high, these approaches are deemed to lead to maladaptation (Barnett & O'Neill 2010). The high opportunity costs of the CDM are embedded in its dual aim of achieving cost-effective GHG emission reductions and sustainable development. As this critique has found, in spite of both objectives being given equal weight (Bozmoski et al. 2008), the opportunity cost of privileging cost-effective mitigation has seen sustainable development side-lined. What this exposes is a broader failure to understand that 'achieving the ultimate goal of the Climate Convention and the Kyoto Protocol is dependent on development paths and socio economic choices at least as much as on climate [mitigation] policy' (Olsen 2007, p. 71). The new mechanism under Article 6.4 of the Paris Agreement shares many features with the CDM, including its dual objective: 'A mechanism to contribute to the mitigation of greenhouse gas emissions and support sustainable development is hereby established ...' (UNFCCC 2015).

As to the mechanism's environmental integrity, the CDM has demonstrated high environmental costs through its failure in leading to additional emission reductions (Muller 2007, cited in Bumpus & Liverman 2008; Cames et al. 2016). As Cames et al. (2016) found in their analysis of 300 project types in the CDM pipeline covering the period 2013–2020, 'only 2% of the projects and 7% of potential CER supply have a high likelihood of ensuring that emission reductions are additional and are not overestimated' (Cames et al. 2016, p. 11). Extrapolating this as a percentage of the more than 5 billion tonnes of CO_2e in Kyoto-era offset units (Evans & Gabbatiss 2019b), 350 million tonnes of CO_2e (7 per cent) in CER supply have a high likelihood that the emission reductions are additional and not overstated.

Environmental integrity as a manifestation of high environmental costs translates as 'no environmental additionality' in the CDM; a result of 'emissions finagling' in both developing countries and in the solid legal systems such as those found in the United Kingdom and United States (Nordhaus 2005, cited in Andrew et al. 2010). This is an enduring limitation of GHG crediting mechanisms, described as an 'inherent and unsolvable dilemma'. That is, on the one hand, perverse incentives are created for host countries and their policy-makers to avoid addressing GHG emissions through the implementation of policies and regulations to do so, as this would lead to a reduction in the potential for international crediting. On the other hand, activities are created that are not additional as they were to be implemented due to policies and regulations in any case (Cames et al. 2016, p. 17).

The theoretical abstraction needed to build a 'fictitious commodity' – propertied pollution rights in the form of CERs generated from CDM projects – to 'credit' against a developed country's binding emission targets under the Kyoto Protocol, freely traded through international emissions trading schemes (Solomon & Heiman 2010), has always been a neoliberal contradiction. The primacy of economic 'efficiency' by its own market logic in seeking elegant

66 *Mitigation and the Kyoto CDM*

solutions to complex problems has, on the evidence presented here, simply failed, and has resulted in squandering time on an approach that is contributing to expediting the planet to the dangerous tipping points of which the science warns the global community. In 2008, NASA physicist James Hansen stated in testimony to the US House Select Committee on Energy Independence & Global Warming that a disturbing conclusion from new research had emerged that shows that the safe level of atmospheric carbon dioxide may only be 350 parts per million (ppm) (Hansen 2008a, 2008b). In 2016, 'the planet's atmosphere broke a startling record: 400 parts per million of carbon dioxide' (Jones 2017). As of 2019, it stands at 409.8 ppm (NOAA Climate.gov 2020).

Determining whether the social, ecological and economic costs are as high for an alternative command and control regulatory regime, which is discussed below, thus requires a closer look at the parameters of the claim that a carbon tax, one of this regime's main policy instruments, is the 'lost opportunity' its advocates purport it to be. The prospect of impending run-away climate change, which occurs when tipping points are breached, is close and may already be breached, as Hansen (2008a, 2008b) reminds us.

The evidence presented suggests that carbon markets to date have legitimised the actual growth of carbon emissions (Bohm & Dabhi 2009). The opportunity cost then of the prevailing dominant market solution to climate change mitigation, fashioned by neoliberal influences to environmental solutions (Andrew et al. 2010), has seen the minimisation of alternative approaches that fall under the broad banner of 'command and control' mechanisms (Solomon & Heiman 2010; Liverman 2009). These instruments to pricing carbon rely on a regulatory approach for their implementation, such as a carbon tax collected through the existing tax infrastructure (Andrew et al. 2010), or measures to ensure 'full-carbon accounting' (Pearce 2008). For example, on 17 July 2014, Australia's brief period of a carbon pricing scheme introduced in July 2012 by the federal Labor government led by Prime Minister Julia Gillard, was subsequently abolished by the federal Liberal National Party led by Prime Minister Tony Abbott two years later, as 'Australia became the first nation in the world to abolish an effective price on emissions' (Mazengarb 2019).

An additional approach to the commodification of mitigation, or what Bumpus and Liverman (2008) describe as 'accumulation by decarbonisation', is to implement an 'inefficient' tax, by theoretical standards, such as a hypothecated or dedicated tax on coal, for example. Prins et al. (2010) assert that rather than relying on the 'sparkling myriad gem' called climate policy embedded in Kyoto's all-inclusive universalism to decarbonise the global energy system, implementing a series of measures requiring direct and ongoing state intervention is needed. For example, implementing a dedicated tax, set at a price that is low enough to be politically feasible, and rising every year thereafter to send a price signal to business. This mechanism alone, however, is insufficient to achieve the emissions reduction needed to avert dangerous climate change and must therefore be supplemented by substantial public-sector

Mitigation and the Kyoto CDM 67

investment in research, development, demonstration and deployment (RDD&D) to act as the main driver for the radical acceleration of decarbonisation as well as to be the major buyer (Prins et al. 2010).

The opportunity cost of abstract market-based mitigation has by contrast served to obscure the transparency of a command and control regulatory approach which has been presented as 'a lost opportunity' when measured against its forgone advantages. That is, relative to emissions trading and off-setting under the CDM, the three main advantages of a carbon tax are: that it is harder to evade or avoid because of its transparency and visibility; the revenue generated by the tax should flow to an accountable government who would then use the extra funds for socially useful purposes; and finally, the state, as the major investor of RDD&D as well as its main buyer, is in the best position to act on the goal of accelerating radical decarbonisation (Andrew 2008, cited in Andrew et al. 2010; Prins et al. 2010).

Proponents of market-based approaches argue this level of state involvement is inefficient. Rarely, however, do these proponents question the inefficiency of added transaction costs for a new global and national governance infrastructure to monitor the market in greenhouse gas emission reductions, let alone acknowledge the state-facilitated expansion of private property rights to bestow 'ownership' on them in the first place. Geographer Diana Liverman (2009) rightfully calls the CDM and its CERs, or 'carbon credits', a 'slippery commodity'. Proof of a counterfactual universe in the CDM always made CERs suspect, and the evidence that actual emissions have increased under the Kyoto-type climate policy has already been presented throughout this chapter.

With respect to the following pathway under which maladaptation arises – that is, how a market-based approach reduces the incentive to adapt – I agree that it is the potential to make profits under this approach, such as emissions trading schemes and the CDM, that is 'the distinguishing feature sought by firms and is at the heart of the neoliberal argument in favour of an ETS [Emissions Trading Scheme] over a carbon tax' (Andrew et al. 2010, p. 614). Viewed through the lens of neoliberal economism (Ralston Saul 2006), it is easy to see why the dual goal of cost effectiveness and sustainable development under the CDM has been difficult to reconcile. There are no potential profits to be made by meeting the goal of sustainable development because, as stated, 'sustainable development has no monetary value in the [CDM] mechanism' (Paulsson 2009, p. 70). Monetary value here is a single exchange value, a social construct, and a blunt metric in determining Payment for Environmental Services (PES) more generally. Therefore, continuing to monetise, through the commodification of environmental services to 'the single language of valuation', is to contribute to making human-nature relationships invisible (Kosoy & Corbera 2010).

In summary, it is apparent that the existing dominant approach to mitigation under the CDM is both inefficient and inequitable. Inefficient due to the high costs associated with the construction of a 'fictitious commodity' in GHG emissions, which in turn seriously diminishes the claim of the CDM's

68 *Mitigation and the Kyoto CDM*

cost-efficient emission-reduction objective. Inequitable, because in aiming for least-cost abatement, sustainable development has been subordinated, thus contributing to the invisibility of human-nature relationships that are vital for the buy-in necessary to mitigate climate change in any socially just, legitimate and environmentally responsible way.

(4) How the CDM reduces incentives to adapt

As the previous pathway has demonstrated, in the absence of alternative approaches to current mitigation policy fashioned by neoliberal environmentalism, social, economic and environmental costs are high, and hence this approach is considered to be maladaptive (Barnett & O'Neill 2010). The opportunity cost of the CDM has been demonstrated through the high social, economic and environmental costs it incurs in developing countries in particular. The greater opportunity cost of global market myopia has been to minimise an alternative regulatory regime implemented in industrialised countries that are more transparent and less in need of their own, contradictory and vast, monitoring infrastructure to give, for example, the CDM's notion of 'additionality' legitimacy.

It is right to claim then, that where new property rights regimes are introduced, 'these are defined by those with economic and social power and, consequently legitimises a particular social order' (Kosoy & Corbera 2010, p. 1234). The social order, like the environmental order under the rubric of market environmentalism, is subsumed within the political order of capitalism, in this case, 'carbon capitalism' (Maxton-Lee 2018, 2020; Osborne 2018) or 'carbon colonialism' (Bachram 2004). Viewed in a more historical light, the new market in carbon must be seen in relation to the development of Marxian capitalism's 'primitive accumulation'. Harvey (2003) substitutes this term for 'accumulation by dispossession', believing it peculiar to call an ongoing process 'primitive' or 'original'. The features of primitive or original accumulation maintain a powerful presence within the historical geography of capitalism and as a continuing force of capital accumulation today. The list of mechanisms of primitive accumulation is extensive and they have been 'fined tuned' to play a greater role now than in the past.[2] For example, Harvey (2003) singles out the credit system and finance capital in particular as the latest iteration of primitive accumulation which, he suggests, 'entails appropriation and co-optation of pre-existing cultural and social achievements as well as confrontation and supersession' (Harvey 2003, p. 146).

Finally, it is important to emphasise and make explicit what has been rendered obscure within the dominant market-based solution to mitigation. That is, how the opportunity costs of this approach's continuation have been obscured by the distinguishing feature sought by capital – to make profit (Delabre et al. 2020; Maxton-Lee 2018; McAfee 2014; Andrew et al. 2010). That is, maladaptation arises under this pathway, when an approach reduces incentives to adapt. In this case, the profit incentive is at the heart of the

Mitigation and the Kyoto CDM 69

neoliberal argument for emissions trading and offsets, and this in turn is what reduces incentives to adapt to an alternative command and control approach.

However, Hansen (2008) provides another metric from another episteme to that of the market. If the scientific research that he and other leading climate scientists suggest is right – and this comes with conclusions that have a certainty exceeding 99 per cent – then 'we have now used up all the slack in the schedule for action needed to defuse the global warming bomb' (Hansen 2008a, 2008b). This is not meant to imply mitigation stops. Biophysical limits have been breached at 400 ppm (Jones 2017). 350 ppm, what the research says is 'the safe level of atmospheric carbon dioxide … and it may be less' (Hansen 2008a, 2008b), requires an approach for radical decarbonisation. Hansen (2008a, 2008b) could not be clearer in advocating a carbon tax to achieve this.

Against a catalogue of evident and multiple failures of the market-based approach to cost-effective mitigation and sustainable development accompanying the 'slippery commodity' of the CDM, is a lexicon of endless variations on invisibility: veils, masks, minimisations, reductionisms – obscurity in general – which appears to accompany this approach (Kosoy & Corbera 2010). Hansen's metric of 350 ppm is clear. Obedience to biophysical limits is thus the departure point (Muradian et al. 2010) for an urgent alternative state regulatory approach that uses command and control mechanisms as they have been presented here.

However, whether the new scientific metric can supplant the powerful primacy of the profit motive metric remains to be seen. My suspicion is it will not, and women will largely be left to pick up what is remaining of the 'slack' as they did so as 'buffers' in the structural adjustment programmes in South America in the 1980s and 1990s (Bessis 2003). More generally, gender issues are treated at UNFCCC COP negotiations as 'being used as fill, rather than as substantive matters for negotiation' (Röhr 2006, p. 8). Illuminating the gender dimension of climate change mitigation is akin to filling the 'social emptiness' (Muradian et al. 2010) of fictitious boundaries created by Kyoto mechanisms. It is the key to recognising that carbon worth has other dimensions than the single language of valuation favoured by market environmentalism. Humannature inter-relationships and interactions are far too complex for neoliberal reductionism in the commodification of environmental services.

The question which might reasonably be asked here is: should funding generated by a 2 per cent levy on CDM projects for the Adaptation Fund, expanded to include Joint Implementation and Emissions Trading in the second commitment period, be directed to a 'Maladaptation Fund' of this ideology's own making?

(5) How market-based mitigation, including the CDM, sets paths that limit the choices available to future generations (path dependency)

The opportunity to 'offset' emissions is bundled up in the lost opportunity of an alternative regulatory regime, as the previous pathways through which

70 *Mitigation and the Kyoto CDM*

maladaptation arises have demonstrated. That is, the CDM has legitimised the growth in actual emissions; operated in the genre of direct foreign investment in favouring the more developed of the developing countries, thus disproportionately burdening the most vulnerable and most in need of adaptation funds; and demonstrated that continuing high opportunity costs are no barrier to the potential to make profits. This, apparently, is what reduces incentives to adapt.

The question addressed here is, does the CDM lead to maladaptation? It is evident the pathways through which maladaptation arises through the CDM are multiple as demonstrated. These pathways are not exhaustive and must be joined by Barnett and O'Neill's (2010) fifth type in which maladaptation arises under their typology. It can be argued that the final pathway through which maladaptation arises is a synthesis of all previous pathways the CDM has exhibited. In summary and in general, the Kyoto flexibility market mechanisms are putting present and future generations on a pathway to maladaptation. In this circumstance, available choices to future generations are circumscribed by biophysical limits that the present generation is in breach of today.

Conclusion

Does the CDM lead to maladaptation? The foregoing analysis suggests that the CDM appears to distinguish itself as a mechanism for manufacturing maladaptation and not mitigation. The market-driven approach to mitigation in the form of the UNFCCC Kyoto Protocol's flexibility mechanisms, reveals that the CDM mitigation policy is the driver of maladaptation where negative consequences in the form of potential displacement, along with inequity, social disparity and injustice, are its companion outcomes.

Chapters 3 and 4 have necessarily questioned the technical and neoliberal dominance of climate change mitigation and adaptation. In the following chapters, I demonstrate how these policies specifically impact on women. That is, by focusing on the gendered nature of maladaptation.

Notes

1 'It is notable that the US or Australia were the only industrialised countries not to ratify the protocol when it went into force without them in 2005. Responsible for 3% (Australia) and 33% (US) of industrialised emissions this reduced the potential effectiveness of the protocol from 5.2% to about 2% (although Australia ratified Kyoto at the end of 2007)' (Liverman 2009, p. 292, footnote 35).

2 'Some of the mechanisms of primitive accumulation ... have been fine tuned to play an even stronger role than in the past. The credit system and finance capital became ... major levers of predation, fraud and thievery ... stock promotions, ponzi schemes, structured asset destruction through inflation, asset stripping through mergers and acquisitions, and the promotion levels of debt incumbency that reduce whole populations to debt peonage ... central features of what contemporary capitalism is all about' (Harvey 2003, p. 147).

References

Adger, W, Paavola, J, Huq, S & Mace, M (eds) 2006, *Toward justice in adaptation to climate change: Fairness in adaptation to climate change*, MIT Press, Cambridge, Massachusetts, pp. 1–19.

Andrew, J, Kaidonis, M & Andrew, B 2010, 'Carbon tax: challenging neoliberal solutions to climate change', *Critical Perspectives on Accounting*, vol. 21, no. 7, pp. 611–618.

Bachram, H 2004, 'Climate fraud and carbon colonialism: the new trade in greenhouse gases', *Capitalism Nature Socialism*, vol. 15, no. 4, pp. 5–20.

Barnett, J & O'Neill, S 2010, 'Maladaptation', *Global Environmental Change*, vol. 20, pp. 211–213.

Bello, W 2009, 'Carbon markets: a fatal illusion', in S Bohm & S Dabhi (eds), *Upsetting the offset: the political economy of carbon markets*, e-book, accessed 21 May 2017, http://mayflybooks.org/wp-content/uploads/2010/07/9781906948078UpsettingtheOffset.pdf.

Bessis, S 2003, 'International organisations and gender: new paradigms and old habits', *Signs: Journal of Women in Culture and Society*, vol. 29, no. 2, pp. 633–647.

Bohm, S & Dabhi, S (eds) 2009, *Upsetting the offset: the political economy of carbon markets*, e-book, accessed 21 May 2017, http://mayflybooks.org/wp-content/uploads/2010/07/9781906948078UpsettingtheOffset.pdf.

Boyd, E, Corbera, E & Estrada, M 2008, 'UNFCCC negotiations (pre-Kyoto to COP-9): what the process says about the politics of CDM-sinks', *International Environmental Agreements: Politics, Law and Economics*, vol. 8, no. 2, pp. 95–112.

Boyd, E, Hultman, N, Roberts, J, Corbera, E, Cole, J, Bozmoski, A, Ebeling, J, Tippman, R, Mann, P, Brown, K & Liverman, D 2009, 'Reforming the CDM for sustainable development: lessons learned and policy futures', *Environmental Science & Policy*, vol. 12, no. 7, pp. 820–831.

Bozmoski, A, Lemos, M & Boyd, E 2008, 'Prosperous negligence: Governing the clean development mechanism for markets and development', *Environment: Science and Policy for Sustainable Development*, vol. 50, no. 3, pp. 18–30.

Bronen, R 2009, 'Forced migration of Alaskan indigenous communities due to climate change: creating a human rights response', in A Oliver-Smith & X Shen (eds), *Linking environmental change, migration and social vulnerability*, UNU Institute for Environment and Human Security (UNU-EHS), e-book, accessed 21 May 2017, www.munichre-foundation.org/dms/MRS/Documents/Source2009_OliverSmith_ShenEnvironmentalChange_Migration.pdf.

Brooks, N, Grist, N & Brown, K 2009, 'Development futures in the context of climate change: challenging the present and learning from the past', *Development Policy Review*, vol. 27, no. 6, pp. 741–765.

Brown, K 2011, 'Sustainable adaptation: an oxymoron?', *Climate and Development*, vol. 3, no. 1, pp. 21–31.

Brown, L 2013, 'Herman Daly Festschrift: restructuring taxes to create an honest market', in J Farley (ed), *The Encyclopedia of Earth*, wiki article, 7 October, accessed 17 June 2017, http://editors.eol.org/eoearth/wiki/Herman_Daly_Festschrift:_Restructuring_taxes_to_create_an_honest_market.

Bumpus, A & Cole, J 2010, 'How can the current CDM deliver sustainable development?', *Wiley Interdisciplinary Reviews: Climate Change*, vol. 1, no. 4, pp. 541–547.

Bumpus, A & Liverman, D 2008, 'Accumulation by decarbonisation and the governance of carbon offsets', *Economic Geography*, vol. 84, no. 2, pp. 127–155.

72 Mitigation and the Kyoto CDM

Camargo, I 2009, 'What's new – July 2009', *Climate Funds Update*, accessed 15 June 2017, www.climatefundsupdate.org/news/whatsnew-july2009.

Cames, M, Harthan, RO, Füssler, J, Lazarus, M, Lee, CM, Erickson, P & Spalding-Fecher, R 2016, *How additional is the Clean Development Mechanism?: analysis of the application of current tools and proposed alternatives*, Study prepared for DG CLIMA, Öko Institut, accessed 7 August 2020, https://ec.europa.eu/clima/sites/clima/files/ets/docs/clean_dev_mechanism_en.pdf.

Castree, N 2006, 'From neoliberalism to neoliberalisation: consolations, confusions, and necessary illusions', *Environment and Planning A*, vol. 38, no. 1, pp. 1–6.

Collett, L 2009, *A fair-weather friend? Australia's relationship with a climate-changed Pacific*, Australia Institute, accessed 21 May 2017, www.tai.org.au/node/1499.

Dabhi, S 2009, 'Where is climate justice in India's first CDM project', in S Bohm & S Siddhartha (eds), *Upsetting the offset: the political economy of carbon markets*, e-book, accessed 21 May 2017, http://mayflybooks.org/wp-content/uploads/2010/07/9781906948078UpsettingtheOffset.pdf.

Daly, H 1995, 'The irrationality of homo economicus', *Developing Ideas Digest*, Developing Ideas interview by Karl Hansen, College Park, Maryland, USA, 8 February, accessed 3 June 2017, http://pratclif.com/sustainability/Herman%20Daly.htm.

Daly, H 2005, 'Living in a finite world: Herman Daly and economics', in CN McDaniel, *Wisdom for a livable planet: the visionary work of Terri Swearingen, Dave Foreman, Wes Jackson, Helena Norberg-Hodge, Werner Fornos, Herman Daly, Stephen Schneider, and David Orr*, Trinity University Press, San Antonio.

Daly, H 2008, 'Special report: economics blind spot is a disaster for the planet', *New Scientist*, vol. 200, no. 2678, pp. 46–47.

Daly, H & Townsend, K 1993, *Valuing the earth: economics, ecology, ethics*, MIT Press.

Delabre, I, Boyd, E, Brockhaus, M, Carton, W, Krause, T, Newell, P, Wong, GY & Zelli, F 2020, 'Unearthing the myths of global sustainable forest governance', *Global Sustainability*, vol. 3, no. 16, pp. 1–10.

Dufrasne, G 2018, 'COP24: Withdrawn UN advert shows why carbon offset scheme should be scrapped', *Carbon Market Watch*, accessed 19 August 2020, https://carbonmarketwatch.org/2018/08/31/withdrawn-un-advert-shows-why-carbon-offset-scheme-should-be-scrapped/.

Elliott, L 2007, 'Improving the global environment: policies, principles and institutions', *Australian Journal of International Affairs*, vol. 61, no. 1, pp. 7–14.

Evans, S & Gabbatiss, J 2018, 'COP24: Key Outcomes agreed at the UN climate talks in Katowice', *Carbon Brief*, accessed 8 August 2020, www.carbonbrief.org/cop25-key-outcomes-agreed-at-the-un-climate-talks-in-katowice.

Evans, S & Gabbatiss, J 2019a, 'In-depth Q & A: how 'Article 6' carbon markets could 'make or break' the Paris Agreement', *Carbon Brief*, accessed 8 August 2020, www.carbonbrief.org/in-depth-q-and-a-how-article-6-carbon-markets-could-make-or-break-the-paris-agreement.

Evans, S & Gabbatiss, J 2019b, 'COP25: Key Outcomes agreed at the UN climate talks in Madrid', *Carbon Brief*, accessed 8 August 2020, www.carbonbrief.org/cop25-key-outcomes-agreed-at-the-un-climate-talks-in-madrid.

Gerebizza, E 2011, 'Can financial markets solve the climate market? unpacking the false solutions by the World Bank', *REDD-Monitor*, accessed 17 June 2017, www.redd-monitor.org/wp-content/uploads/2011/03/Discussion_note-march2011.pdf.

Mitigation and the Kyoto CDM   73

Goldtooth, T 2010, 'Why REDD/REDD+ is not a solution', in J Cabello & T Gilbertson (eds), *No REDD!: a reader*, e-book, accessed 15 June 2017, http://wrm.org.uy/wp-content/uploads/2013/04/REDDreaderEN.pdf.

Hansen, J 2008a, 'Twenty years later: tipping points near on global warming', *Huffpost* 1999, weblog post, 1 July, accessed 17 June 2017, www.huffingtonpost.com/dr-james-hansen/twenty-years-later-tippin_b_108766.html.

Hansen, J 2008b, 'Global warming twenty years later: tipping points near', *Yale School of Forestry & Environmental Studies, The Forum on Religion & Ecology at Yale*, accessed 10 June 2017, www.columbia.edu/~jeh1/2008/TwentyYearsLater_20080623.pdf.

Harvey, D 2003, *The new imperialism*, Oxford University Press Inc., New York, pp. 137–182.

Harvey, D 2005, *A brief history of neoliberalism*, Oxford University Press Inc., New York.

Harvey, D 2007, 'Neoliberalism as creative destruction', *Annals of the American Academy of Political and Social Science*, vol. 610, pp. 22–44.

Hofbauer, J 2015, *Climate-Policy Induced Migration (January 17, 2015)*, ClimAccount Working Paper Series No. KR13AC6K11043, accessed 25 February 2018, https://ssrn.com/abstract=2797166.

Jones, N 2017, *How the world passed a carbon threshold and why it matters*, Yale Environment 360, accessed 6 June 2017, http://e360.yale.edu/features/how-the-world-passed-a-carbon-threshold-400ppm-and-why-it-matters. Yale School of Forestry & Environmental Studies.

Juhola, S, Glaas, E, Linner, B & Neset, T 2016, 'Redefining maladaptation', *Environmental Science & Policy*, vol. 55, January, pp. 135–140.

Kolmannskog, V 2009, *Climate change, disaster, displacement and migration: initial evidence from Africa*, UNHCR, The UN Refugee Agency Policy Development and Evaluation Service, accessed 15 June 2017, www.unhcr.org/4b18e3599.html.

Kosoy, N & Corbera, E 2010, 'Payments for ecosystem services as commodity fetishism', *Ecological Economics*, vol. 69, no. 6, pp. 1228–1236.

Liverman, D 2009, 'Conventions of climate change: constructions of danger and the dispossession of the atmosphere', *Journal of Historical Geography*, vol. 35, no. 2, pp. 279–296.

Lohmann, L 2009a, 'Regulation as corruption in the carbon offset markets', in S Bohm & S Siddhartha (eds), *Upsetting the offset: the political economy of carbon markets*, e-book, accessed 21 May 2017, http://mayflybooks.org/wp-content/uploads/2010/07/9781906948078UpsettingtheOffset.pdf.

Lohmann, L 2009b, 'Regulation as corruption in the carbon offset markets: cowboys and choirboys united', *The Corner House*, accessed 10 June 2017, www.thecornerhouse.org.uk/sites/thecornerhouse.org.uk/files/Athens%2010.pdf.

Magnan, AK, Schipper, ELF, Burkett, M, Bharwani, S, Burton, I, Erikson, S, Gemenne, F, Schaar, J & Zievogel, G 2016, 'Addressing the risk of maladaptation to climate change', *Wiley Interdisciplinary Reviews: Climate Change*, vol. 7, September/October, pp. 646–665.

Mathur, V, Afionis, S, Paavola, J, Dougill, A & Stringer, L 2014, 'Experiences of host communities with carbon market projects: towards multi-level climate justice', *Climate Policy*, vol. 14, no. 1, pp. 42–62.

Maxton-Lee, B 2018, 'Narratives of sustainability: a lesson from Indonesia: global institutions are seeking to shape an understanding of sustainability that undermines

74 Mitigation and the Kyoto CDM

its challenge to their world view', *Sounding: A journal of politics and culture*, no. 70, Winter, pp. 45–57.

Maxton-Lee, B 2020, 'Forests, carbon markets, and capitalism: how deforestation in Indonesia became a geo-political hornet's nest', Guest Post, *REDD-Monitor*, accessed 22 August 2020, https://redd-monitor.org/2020/08/21/guest-post-forests-ca rbon-markets-and-capitalism-how-deforestation-in-indonesia-became-a-geo-political-hornets-nest/.

Mazengarb, M 2019, 'Five years after carbon price repeal, Australia remains in policy abyss', *Renew Economy*, accessed 20 August 2020, https://reneweconomy.com.au/five-years-after-carbon-price-repeal-australia-remains-in-policy-abyss-43066/.

McAdam, J 2007, 'Climate change "refugees" and international law', *NSW Bar Association*, 24 October.

McAdam, J 2012, *Climate change, forced migration and international law*, Oxford University Press, Oxford.

McAfee, K 1999, 'Selling nature to save it? biodiversity and green developmentalism', *Environment and Planning D: Society and Space*, vol. 17, pp. 133–154.

McAfee, K 2014, *Green economy or buen vivir: can capitalism save itself*, Lunchtime Colloquium, Rachel Carson Centre for Environment and Society, online video, accessed 17 May 2020, www.carsoncenter.unimuenchen.de/events_conf_seminars/event_history/2014-events/2014_lc/index.html.

McCarthy, J & Prudham, S 2004, 'Neoliberal nature and the nature of neoliberalism', *Geoforum*, vol. 35, no. 3, pp. 275–283.

McMurray, D, Hall-Taylor, B & Porter, E 2002, *Unit EDU40001 research methods in social sciences: study guide*, School of Social and Workplace Development, Southern Cross University, Lismore, Australia.

Muradian, R, Corbera, E, Pascual, U, Kosoy, N & May, P 2010, 'Reconciling theory and practice: an alternative conceptual framework for understanding payments for environmental services', *Ecological Economics*, vol. 69, no. 6, pp. 1202–1208.

Najam, A, Huq, S & Sokona, Y 2003, 'Climate negotiations beyond Kyoto: developing countries concerns and interests', *Climate Policy*, vol. 3, no. 3, pp. 221–231.

National Oceanic and Atmospheric Administration (NOAA) Climate.gov 2020, *Climate change: atmospheric carbon dioxide*, NOAA Climate.gov, accessed 20 August 2020, www.climate.gov.

Oliver-Smith, A & Shen, X (eds) 2009, *Linking environmental change, migration and social vulnerability*, UNU Institute for Environment and Human Security (UNU-EHS), e-book, accessed 21 May 2017, www.munichre-foundation.org/dms/MRS/Docum ents/Source2009_OliverSmith_ShenEnvironmentalChange_Migration.pdf.

Olsen, K 2007, 'The clean development mechanisms contribution to sustainable development: a review of the literature', *Climatic Change*, vol. 84, no. 1, pp. 59–73.

Osborne, T 2018, 'The de-commodification of nature: indigenous territorial claims as a challenge to carbon capitalism', *Environment and Planning E: Nature and Space*, vol. 1, no. 1–2, pp. 25–75.

Paulsson, E 2009, 'A review of the CDM literature: from fine-tuning to critical scrutiny?', *International Environmental Agreements: Politics, Law and Economics*, vol. 9, no. 1, pp. 63–80.

Pearce, F 2008, 'Carbon trading: dirty, sexy money', *New Scientist*, vol. 198, no. 2562, pp. 38–41.

Pielke, R, Prins, G, Rayner, S & Sarewitz, D 2007, 'Climate change 2007: lifting the taboo on adaptation', *Nature*, vol. 445, no. 7128, pp. 597–598.

Prins, G, Galiana, I, Green, C, Grundmann, R, Hulme, M, Korhola, A, Laird, F, Nordhaus, T, Pielke, R, Rayner, S, Sarewitz, D, Shellenberger, M, Stehr, N & Tezuka, H 2010, *The Hartwell papers: a new direction for climate policy after the crash of 2009*, London School of Economics, accessed 21 May 2017, http://eprints.lse.ac.uk/27939/1/HartwellPaper_English_version.pdf.

Prins, G & Rayner, S 2007, 'Time to ditch Kyoto', *Nature*, vol. 449, no. 7165, pp. 973–975.

Ralston Saul, J 2006, *The collapse of globalism: and the reinvention of the world*, Penguin Books, Camberwell, Victoria.

Röhr, U 2006, 'Gender relations in international climate change negotiations', *Berlin: LIFE eV/genanet.*

Schipper, ELF 2006, 'Conceptual history of adaptation in the UNFCCC process', *Review of European Community & International Environmental Law*, vol. 15, no. 1, pp. 82–92.

Silvey, R 2009, 'Development and geography: anxious times, anaemic geographies, and migration', *Progress in Human Geography*, vol. 33, no. 4, pp. 507–515.

Smith, K 2008, 'Offset standard is off target', *The Corner House*, accessed 2017, www.thecornerhouse.org.uk/resource/offset-standard-target.

Smith, K 2009, 'Offset under Kyoto: a dirty deal for the south', in S Bohm & S Siddhartha (eds), *Upsetting the offset: the political economy of carbon markets*, e-book, accessed 21 May 2017, http://mayflybooks.org/wp-content/uploads/2010/07/9781906948078UpsettingtheOffset.pdf.

Solomon, B & Heiman, M 2010, 'Integrity of the emerging global markets in greenhouse gases', *Annals of the Association of American Geographers*, vol. 100, no. 4, pp. 973–982.

Stewart, MA 2010, 'Swapping air, trading places: carbon exchange, climate change policy and naturalizing markets', *Radical History Review*, vol. 2010, no. 107, pp. 25–43.

Sutter, C & Parreno, J 2007, 'Does the current clean development mechanism (CDM) deliver its sustainable development claim? An analysis of officially registered CDM projects', *Climate Change*, vol. 84, no. 1, pp. 75–90.

United Nations Environment Programme (UNEP)-Risoe Centre CDM Pipeline 2011, last updated 1 March 2011, accessed 30 March 2011, http://cdmpipeline.org/regions_7.html.

United Nations Framework Convention on Climate Change (UNFCCC) 1992, United Nations General Assembly, 20 January 1994, accessed 9 May 2017, www.refworld.org/docid/3b00f2770.html.

United Nations Framework Convention on Climate Change (UNFCCC) 1998, *Kyoto Protocol to the United Nations Framework Convention on Climate Change*, United Nations, accessed 20 May 2017, https://unfccc.int/resource/docs/convkp/kp eng.pdf.

United Nations Framework Convention on Climate Change (UNFCCC) 2015, *Paris Agreement*, accessed 16 September 2020, https://unfccc.int/process-and-meetings/the-paris-agreement/the-paris-agreement.

United Nations Framework Convention on Climate Change (UNFCCC) Adaptation Fund 2011, *Adaptation Fund*, United Nations Framework Convention on Climate Change, accessed 30 June 2011, http://unfccc.int/kyoto_protocol/mechanisms/clean_development_mechanism/items/2718.php.

United Nations Framework Convention on Climate Change (UNFCCC) CDM 2011, *Clean Development Mechanism (CDM)*, United Nations Framework on Climate

Change, accessed 30 June 2011, http://unfccc.int/kyoto_protocol/mechanisms/clean_development_mechanism/items/2718.php.

United Nations Framework Convention on Climate Change (UNFCCC) Conference of Parties (COP) 16 AWG-LCA 2010, *Report on the Conference of Parties on its sixteenth session, held in Cancun from 29 November to 10 December*, United Nations Framework Convention on Climate Change, accessed 15 June 2017, http://unfccc.int/resource/docs/2010/cop16/eng/07a01.pdf#page=2.

Wara, M 2007, 'Is the global carbon market working?', *Nature*, vol. 445, no. 7128, pp. 595–596.

Work, C, Rong, V, Song, D & Scheidel, A 2019, 'Maladaptation and development as usual? Investigating climate change mitigation and adaptation projects in Cambodia', *Climate Policy*, vol. 19, no. 51, pp. 547–562.

Wysham, D 2008, 'Carbon market fundamentalism', *Multinational Monitor*, vol. 29, no. 3, accessed 15 May 2017, www.multinationalmonitor.org/mm2008/112008/wysham.html.

5 'Silent offsets' and feminist perspectives on women, climate change, UN-REDD+

Adapting to women

Introduction

This chapter takes us from the general to the theoretical, with a particular focus on women in the United Nations Programme on Reducing Emissions from Deforestation and Forest Degradation (UN-REDD Programme). Two aims will be covered over two broad sections.

In the first section, the aim is to provide an in-depth background commentary explaining and linking the main themes and concepts of the chapter, namely: women as 'silent offsets' and REDD+, a United Nations collaborative initiative. On a theoretical level, the broad thematic content is situated in the spatial messiness of global, national and local frameworks. These frames are understood as both the theoretical and policy responses to climate change via mitigation and adaptation policies that have veered to new 'enclosure' processes of commodification central to, for example, the United Nations Framework Convention on Climate Change (UNFCCC) Kyoto Protocol market mechanisms of emissions trading and Clean Development Mechanism (CDM) offsets.

The early silence/invisibility of women in climate change discourse more generally, opens the analysis in section 1, thus beginning the process of theorising the silence surrounding women. REDD+ is the latest mechanism to join the global carbon trading suite, having been recognised as a valid mitigation strategy in the 2009 Copenhagen Accord (Mustalahti et al. 2012). Thus, a broad sweep of the socio-political-legal-economic and environmental landscape of this chapter is interwoven and mapped throughout. Emergent concepts, such as women as 'silent offsets' in developing countries, are conceptualised through interdisciplinary critiques spanning feminisms, neoliberal economics, economic geography and ecological economics to foster a deeper understanding, thus interpretation of what is meant by this concept.

Section 2 of the chapter aims to link theoretical insights gained in the first section on women as silent offsets in the global South, through an analysis on operationalising REDD+ by drawing on insights from the metaphors, strategies and politics of 'feminising the economy' (Cameron & Gibson-Graham 2003), to define what a silent offset is. REDD+ countries are in developing

78 'Silent offsets' and feminist perspectives

countries across the tropics, primarily targeting countries with high forest cover and rates of deforestation (Phelps et al. 2010). Drawing on early studies from REDD+ pilot and participating countries (World Rainforest Movement 2010; Mustalahti et al. 2012), the REDD+ regime exhibits a dangerous pathway to maladaptation with the potential for displacement of forest peoples from their extended livelihood lands in developing countries host to REDD+ projects and national-level REDD+ programmes. The potential for this outcome is further buttressed in a report released by the UN-REDD Programme by 'calling attention to the dangers of REDD marginalising the landless and potentially causing conflict, forced relocation and displacement of forest peoples' (UN-REDD Programme n.d., cited in Goldtooth 2009).

To avoid these unjust outcomes, the UN-REDD Programme claims that 'at a minimum, REDD+ programmes and policies must adhere to the "do no harm" principle. As a baseline assumption, perpetuation or exacerbation of existing inequalities constitutes harm' (UN-REDD Programme 2011, p. 12). This assumption inheres the precautionary principle (PP). In operationalising REDD + then, the PP should serve, at least theoretically, as a guide to adhering to the do no harm principle, thus avoiding known pathways to maladaptation (Barnett & O'Neill 2010).

REDD+ has the ambiguous potential 'to be one of the most significant changes to forest management approaches' (Dooley 2008, cited in Phelps et al. 2010, p. 322), or 'the biggest land grab of all time within the forested lands of Indigenous Peoples' (Goldtooth 2009). The potential negative social, economic and environmental impacts are gendered and radically altered by the gender silencing of women in the policy of REDD+ and its implementation. Women in REDD+ developing countries, indigenous women in particular, are disproportionately affected, not only by the effects of climate change where 'they are different and more severe for women because of their socially allocated role' (Filippini 2011), but they also face limitations when it comes to participating in policy responses to climate change (Gender CC 2011) in the form of REDD+. The potential and existing impacts of climate change and its solutions therefore double their climate change jeopardy in spite of promises of a triple win delivering forest conservation, poverty alleviation and climate mitigation (Delabre et al. 2020).

Women of the global South have been obscured by what McAfee (2012) refers to as the 'asocial logic of neoclassical economics' framing REDD+ projects, which fall more broadly under the umbrella of Payment for Environmental Services (PES) schemes and policies. REDD+ is an offset mechanism that exhibits all the hallmarks of the five-pathway maladaptive trajectory laid out in the previous chapter on the CDM through Barnett and O'Neill's (2010) maladaptation typology. Nevertheless, the World Bank cites the CDM 'as the main model for REDD and ES [environmental services] as a climate policy strategy' (McAfee 2012, p. 112), to be 'financed by carbon-offset trading' (McAfee 2012, p. 105). That is, 'REDD+ is basically PES on steroids' (McAfee 2014).

The criticisms *within* this critique of REDD+, centre less on noble intentions of effective mitigation of emissions from deforestation and forest degradation in developing countries and other co-benefits (denoted by the plus sign). That is, socioeconomic co-benefits such as the protection of human rights, improved forest governance, pro-poor development. And environmental co-benefits, such as quality and provision of water and soil, sustainable forest management, biodiversity protection (Mustalahti et al. 2012, p. 16). Although co-benefits may be offered by REDD+, such as land tenure and social development, 'these co-benefits are not guaranteed' (Chhatr & Agrawal 2009, cited in Mustalahti et al. 2012, p. 16). Rather, the following criticisms of REDD+ centre on this mitigation offset's potential to lead to maladaptation, presenting as a different driver to the displacement of forest-dependent peoples in which the projects are implemented.

UN-REDD+ is premised on a neoliberal economic logic that revolves around abstractions – for example, 'additionality' and counterfactuals such as offsets. 'Offsets are an imaginary commodity created by deducting what you hope happens from what you guess would have happened' (Welch 2007, cited in Lang 2012a). Without the counterfactual (non-existent) version of events, it is impossible to know how many tonnes of carbon were not emitted into the atmosphere, thus impossible to know how many carbon credits to issue from a REDD project at the national level of the REDD Programme (Lang 2012a). The forest is no longer a means of extended livelihood but an emission reduction unit that is traded, securitised and made a derivative of a global carbon market (Lohmann 2011b). Other critics of REDD+ observe that 'REDD dreams … are likely to become REDD nightmares in reality' (Lovera 2009, p. 47). This chapter illuminates what this may mean for women in REDD+ projects, later and specifically. For now, section 1 begins by an analysis on 'a stranger silence still' (MacGregor 2009), the silence/invisibility of women in climate change academic discourse more generally.

(1) Background: de-gendered economics – from market economics to market society

If adaptation as a policy response to climate change through its subordination to mitigation policy was historically biased, then arguably a meaningful participative gender/women dimension in both climate change responses has been flattened altogether. Although this has somewhat changed, particularly in the last decade since the Conference of Parties (COP) 13 in Bali in 2007, the historical roots of this omission, 'gender gap' or 'missing link' are, in part, attributable to the false notion that climate change is a gender-neutral issue, subscribed to by the Intergovernmental Panel on Climate Change (IPCC), through to the UNFCCC and Kyoto Protocol (Röhr 2006). Unsurprisingly then, the most written about aspect in the limited literature on women/gender and climate change, as MacGregor (2009) observed, was the dearth of academic literature in the first place (with a few notable exceptions in the gender

80 *'Silent offsets' and feminist perspectives*

and development scholarship). A critical look at the silence on gender/women attempts to explain why this is so. For example, viewing climate protection as gender neutral neatly dovetails the feminist school of economics situated within the dominant neoclassical liberal paradigm. Feminists of the neoclassical school view the 'market as sexually neutral' (Bessis 2003, p. 635). However, there is nothing asexual about carbon markets when tethered to REDD+ with consequent impacts, in particular, on indigenous communities and women in the creation of a carbon bubble of 'silent offsets'.

Neoliberal economics itself is increasingly under scrutiny for its theoretically myopic world view. At a conference in the United Kingdom focusing on the current international banking system and sustainability, in the context and following on from the 2008 global financial crash, ecological economist Mary Mellor (2012) rejected a claim made by one of the previous speakers who had claimed that academic economics is pluralistic. Mellor suggested that the reason for the 'Just Banking conference' at all, was precisely because academic economics is not pluralistic. Having long refused to discuss alternative models or views, views that challenge the current model of capitalist patriarchal economics, 'where we haven't been able to have these [feminist ecological economic] views heard for 30 years' (Mellor 2012), academic economics has no legitimacy in claiming pluralism. A more accurate claim lies in the illegitimacy of its apparent inevitability, doggedly promoted over the neoliberal decades (1980–2020). Sandel (2013) has suggested that without quite realising it, we have moved from a market economy to a market society.

Specifically, this chapter focuses on the impacts of the UN-REDD Programme on indigenous and forest-dwelling peoples and women in particular, in developing countries; countries that are referred to in crude neoliberal market language as 'low-hanging fruit' (Redman 2008). This term refers to countries of the equatorial, sub-tropical and tropical forests of the globe, particularly those with high deforestation rates (Redman 2008).

Under UN-REDD – used interchangeably with REDD+, where the plus sign refers to environmental co-benefits of conservation, sustainable management of forests and enhancement of forest carbon stocks (UN-REDD Programme 2011) – the communities who are dependent on the forests for their livelihoods and well-being are subsumed in a process described by Bumpus and Liverman (2008) as 'dispossession by decarbonisation'. This is a process that the UNFCCC appears committed to, albeit described in different terms, such as incentivising countries to reduce deforestation. That is, paying governments in developing countries, for example, in the Congo Basin – Cameroon, Central African Republic, Republic of the Congo, Democratic Republic of the Congo, Equatorial Guinea, Gabon (World Rainforest Movement 2010) – as well as countries in the Amazon Basin and Asia-Pacific, to keep their remaining forests intact (Karsenty & Ongolo 2011). The reason for this early exploration of UN-REDD in this background to this first section is because REDD+ is a key concept of the whole chapter and deserves early attention to link and contextualise the chapter's central focus: women and their interaction with REDD;

a mitigation mechanism nestled and negotiated within a global institutional framework that bears close critical analysis.

> The originality of the REDD proposal is its incentive-based mechanism designed to reward the governments of developing countries for their performance in reducing deforestation as measured against a baseline. This mechanism is founded on the hypothesis that developing countries 'pay' an opportunity cost to conserve their forests and would prefer other choices and convert their wooden lands to other uses.
>
> (Karsenty & Ongolo 2011, p. 1)

Karsenty and Ongolo (2011) rightly question the appropriateness of using a theory of incentives with respect to the REDD+ mechanism in 'fragile states' to reduce deforestation. Central to the theory of incentives is the notion of 'principal-agent' where it is assumed that governments as economic agents behave rationally in making decisions to alter their development trajectory. Decisions such as relative cost analysis of alternatives, followed by action and implementation of the rational choice arrived at, thereby facilitating reductions in deforestation. However, such an approach ignores the political economy of the state, especially 'fragile' states, such as 'the Democratic Republic of Congo (the archetype of a fragile state)' (Karsenty & Ongolo 2011, p. 3). Thus, the assumptions underpinning the REDD approach, Karsenty and Ongolo (2011) argue, deserve critical attention.

Firstly, REDD+ assumes governments of these states are in a position to make a decision, based on a cost-benefit analysis that promises financial incentives, to shift their development pathway. Secondly, it is further assumed that once governments in these 'failed' states have made the decision, 'it is capable, thanks to the financial rewards, to *implement and enforce the appropriate policies and measures* which could translate into deforestation reduction' (Karsenty & Ongolo 2011, p. 1, italics in original). Counterfactuals and hypothetical scenarios, measures of greenhouse gas (GHG) emissions against 'baselines' for 'additionality' to be established, are deeply problematic as has been demonstrated elsewhere (see Chapter 4). The UN-REDD Programme promotes the same level of abstraction. Undeterred by those opposed to 'selling nature to save it' (McAfee 1999, 2012), the World Bank cites 'the CDM as the main model for REDD' in climate policy strategy (World Bank 2010a, cited in McAfee 2012, p. 112).

UN-REDD was launched in 2008 as a collaborative initiative built on the technical expertise and convening role of the United Nations Development Programme (UNDP), the Food and Agricultural Organization of the United Nations (FAO) and the United Nations Environment Programme (UNEP) (UN-REDD Programme 2011). Working in tandem with the UNFCCC in Bali in 2007, the World Bank launched its Forest Carbon Partnership Facility (FCPF) (Gilbertson 2010), declaring that UN-REDD and the World Bank (WB) are committed to working 'together both at the international level,

82 'Silent offsets' and feminist perspectives

harmonising normative frameworks and organising joint events, and at the national level, where joint missions and sharing of information are producing coordinated support interventions' (UN-REDD n.d., cited in World Rainforest Movement 2010, p. 4). In other words, the FCPF is the main mechanism through which the WB promotes REDD+. In launching the FCPF, the senior natural resources management specialist at the World Bank, Benoit Bosquet, stated, 'the facility's ultimate goal is to jump-start a forest carbon market that tips the economic balance in favour of conserving forests' (Bosquet 2007, cited in Lang 2009, p. 215). The Indigenous Environmental Network (IEN), who represent native peoples of the Americas and are heavily involved in advocacy and community mobilisation on issues of environmental justice, have suggested that believing and acting as if 'REDD/REDD+ might someday be financed by a repayment of the ecological debt the North owes the South, or by a benevolent fund using public or non-market donations, could be naïve' (Goldtooth 2010).

Herman Daly (1995) observed that most economists working at the WB are protégés of the academic departments of economics from the top-rated universities across the world including, MIT, Oxford, McGill and Harvard, which all teach much the same orthodox economics. '[T]he World Bank is populated by the products of these places, they're eager to receive the advice from them, and I think that's a fundamental problem' (Daly 1995). What the advice from orthodox neoliberal economics has offered thus far is: assumed rational principal-agent roles, notwithstanding the problem of this application in the context of 'fragile' and sometimes 'failed' states (Karsenty & Ongolo 2011); markets that are allegedly sexually neutral to achieve 'liberal' sexual equality (Bessis 2003); markets that are constructed from the alleged social vacuity of 'low-hanging fruit' in the developing world's forests under REDD+; and the minimisation, indeed silencing, of alternative economic views that give salience to women in particular (Mellor 2012). Neoliberal prescriptions are thus founded on myth, where 'myths draw their power from belief, not facts ... a taken-for granted logic which is beyond questioning' (Essebo 2013, p. 6, cited in Delabre et al. 2020, p. 2).

It has been a decade and a half since Stern (2006) declared climate change as 'a major market failure' (Stern 2006, p. 1), nevertheless advocated for market-based solutions to address it. Dissenting voices warned that tackling the climate change problem should not be left to 'the irrationalities and potentially speculative exuberance of an emboldened global carbon market to any large degree' (Schalatek et al. 2010, p. 4). Notwithstanding, this is what Stern (2006, 2009) has consistently argued in favour of. The 'analysis' of climate change action in his book *The Global Deal* (2009), in the end, says the numbers are 'not to be "taken too seriously", as all numbers are "very sensitive to assumptions", "leave out conflict" and are "weak on risk and biodiversity"' (Stern 2009, p. 141, cited in Lohmann 2011a, p. 654).

Illusion and myth are not monopolised by neoclassical economics and extend to certain tenets of arch economic rival, ecofeminism – the marginalised strand of feminism (Kuletz 1992). Indeed, the legacy of ecofeminism's 'radical

'Silent offsets' and feminist perspectives 83

environmental myths' (Jackson 1995) may offer, in part, an explanation for the silence of women in climate change discourse, only if it holds 'that the low status of ecofeminism is partially to blame for the avoidance of environmental issues by mainstream feminist scholars [in general]' (Banerjee & Bell 2007, cited in MacGregor 2009, p. 126). Ecofeminism's 'heyday' locates its emergence in the mid-to-late 1970s (Kuletz 1992; Mellor 1997), through to the early to mid-1990s before an assault from academics questioned, among others ideas, the key ecofeminist notion that women, by virtue of being women, are inherently closer to nature than are men (Leach 2007). In short, the ideas from ecofeminism gained some instrumental traction in putting women and the environment at least on the development agenda, in these early years. However, the resulting development practice of this view saw women in developing countries being burdened with additional unpaid labour on an environmental development project. This was commonly done in the name of some assumed environmental saviour status, due to a false notion that women are especially 'close to nature' in a conceptual or spiritual sense (Leach 2007; Jackson 1993, 1995). Essentialism, however, as acknowledged by its own theorists, 'is the core problem of ecofeminism' (Holland-Cunz 1992, cited in Kuletz 1992).

However, the importance of a set of ideas that took root over 40 years ago as being, in part, responsible for the contemporaneous poverty of feminist engagement with environmental issues cannot be ignored, and is only relevant insofar as providing a partial explanation for the dearth of academic literature on women, gender and climate change. The fact that this book has chosen to use the word 'women' in the title of this chapter may be controversial for some, because as Jackson (1993) claims, when writers of women, environment and development (WED), previously women in development (WID) (Leach 2007), limit their analysis to 'women, an essentialism, is expressed in not only the failure to disaggregate the category "women" but also the failure to realise that women are gendered in relation to men' (Jackson 1993, p. 1950). While there is no disagreement here, the broader concern is that the '[f]requent use of the term *gender* often masks a conceptual vacuum' (Bessis 2003, p. 636, italics in original). As one international development scholar, feminist practitioner and activist reflected, 'we fought so long to have the word women included at all' (E Pittaway 2012, pers. comm., 30 Apr.).

Contrary to Leach's (2007) early tentative conclusion that ecofeminism's 'day has passed', the chapter argues that this remains to be seen. Although Leach (2007) did not name 'ecofeminism' specifically, her intention was clear by reference to the discredited 'discourse on natural environmental carer'. However, what needs to be acknowledged following the two decades of painful and vitriolic exchanges from ecofeminist critics such as Jackson (1993, 1995) and Leach (2007) towards ecofeminism's 'fables' and 'myths', is their own absolutism to the 'essentialism question' (Kuletz 1992). This intradisciplinary debate has blunted the wider scholarship of ecofeminist thought, and the endeavours by its proponents to transcend this deep flaw with concepts such as 'the simultaneous' (Holland-Cunz 1992, cited in Kuletz 1992), 20-year-old

84 'Silent offsets' and feminist perspectives

truncations of a wide body of ecofeminist literature was itself totalising. That is, ecofeminism's critics obscured other significant insights into the political economy of global relations made by ecofeminists and the more relevant contribution this sub-discipline made and can make in its critique of orthodox economics in particular (Mellor 1992, 1997, 2012).

Paralleling ecofeminism's emergence in the mid- to late 1970s, the seeds of contemporary economic globalisation, neoliberalism, were firmly planted with the political arrivals of Margaret Thatcher in the United Kingdom in 1979 and Ronald Reagan in the USA in 1980. Four decades later, one might reasonably ask why neoliberalism thrived and ecofeminist economic views diminished. Views discussed later, that are embodied in the metaphors of Hazel Henderson's (1982) 'Layer Cake with Icing', depicting the Total Productive System of an Industrial Society; Bennholdt-Thomsen and Maria Mies's (1999) 'Iceberg Model of Capitalist Patriarchal Economics'; and Mary Mellor's (2012) 'Economic Dualism'.

Several years after the Global Financial Crisis (GFC) of 2008, Marxian geographers Doreen Massey and David Harvey were co-panellists at a post-GFC Spatial Justice Workshop at the University of Westminster in 2011. Massey (2011) declared the reason for neoliberalism's triumphalism is because 'the Right have won the battle of the narratives', 'the ideological battle'. Further, the contemporary global economy may be in crisis, though differentially experienced across the global geography, it is still a global economic crisis (Harvey 2011; Massey 2011). The crisis, however, appears insufficient for there to be a shift in the balance of social forces. For that to occur, Massey (2011) claimed, there must be a 'conjunctural crisis' – that is, when political, ecological, cultural and economic forces are simultaneously in 'a questioning moment'. Whether the global COVID-19 pandemic crisis of 2020 will be sufficient to shift the balance of social forces towards a conjunctural crisis is a moment in progress; the GFC, which Stuart Hall claims is the climate crisis and sixth mass extinction crisis each on their own and together, failed to do so and is discussed further in the final chapter of this book. Following Massey, David Harvey (2011) observed 'it is undeniable that the processes that are changing this world are the processes of capital accumulation. Everybody does try to play around the fringes of it'. REDD+, a retrofit, is a new site and mechanism for capital accumulation. It was neither designed, nor created to grant money to indigenous or traditional forest communities for the conservation of forests that they already undertake (World Rainforest Movement 2008). Rather, as the World Rainforest Movement's contribution to the Convention on Climate Change in November 2008 stated, '[The REDD] objective is to reduce emissions from deforestation. That implies a scenario where, unless money is made available, the forest will be destroyed' (World Rainforest Movement 2008, p. 5). The funds for market-based conservation schemes are contingent on for-profit investment, thus reinforcing existing 'North-South and South-South inequalities and, will probably create new ones' (McAfee 2014).

A brief historical overview of ecofeminism

Ecofeminism emerged at a time which was synchronous with the beginning of ecological awareness and second wave feminism at its height (Holland-Cunz 1992, cited in Kuletz 1992). However, it was neoliberalism's ascent that prevailed. Where 'its set of ideas have been drawn out and used' (Massey 2011) over the last 40 years, ecofeminism 'withered on the academic vine' plagued 'by a negative reputation as being spiritualist, essentialist, and downright "fluffy"' (MacGregor 2009). Much of the criticism is reserved for Maria Mies and Vandana Shiva's 1993 publication of Ecofeminism (Mellor 1996). However, as Mellor observed,

> like many ecofeminist texts it is part thesis and part treatise, and is therefore easily criticised for hyperbole and unwarranted generalisations. The same might be said of the Communist Manifesto, but movements and ideas have to start somewhere and both radical environmentalism and ecofeminism are still fairly young.
>
> (Mellor 1996, p. 132)

'Fairly young', for the ecofeminist movement did not reach maturation like neoliberalism and that is because, as Massey (2011) previously asserted, the set of ideas that animated the latter doctrine, the dominant forces of global capitalism driving them, had won the battle of the narratives. Scholarly obsessions about ecofeminism's essentialism question had the effect of obscuring other important contributions of conceptualising 'economy', such as the 'Layer Cake with Icing' (Henderson 1982; Figure 5.1) and the 'Iceberg Model' (Bennholdt-Thomsen & Mies 1999; Figure 5.2). Both are metaphors for the 'whole' economy, illustrating 'how so-called productive economic activities [the icing and the tip] depend on a set of currently invisible processes' (Cameron & Gibson-Graham 2003, p. 8). Both appear, however, not to have captured the broad ideological imagination of 'all-ships-rising neoliberalism'.

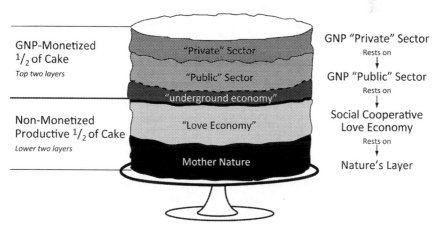

Figure 5.1 Total productive system of an industrial society: Layer Cake with Icing
Source: Henderson 1982 (used with permission).

86 *'Silent offsets' and feminist perspectives*

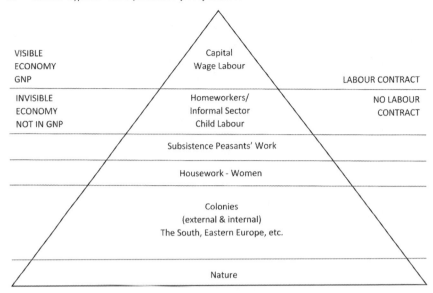

Figure 5.2 The Iceberg Model of capitalist patriarchal economics
Source: Bennholdt-Thomsen and Mies 1999, p. 31 (used with permission).

The 'silent offset' economy of women in REDD+ in developing countries should be added to the bulk of these conceptualisations. Mary Mellor's (2012) 'Economic Dualism' (Figure 5.3), a later iteration on these earlier models, one that dispenses with the visual hierarchy, is used to conceptually frame the construction and analysis of what the 'silent offset' economy means.

Mellor (2012) prefers the use of the term 'provisioning' rather than 'economy'. That is, where 'economics' is taken to mean a 'system of provisioning for the *whole* human community that is sufficient to meet survival and quality of life needs, while being ecologically sustainable. A provisioning that ... [is a] "sustaining economy"' (Mellor 1992, p. 134, italics in original). Provisioning thus involves integrating the despised other half of the economic dualism, where mostly women reside and nature is externalised (Mellor 1997).

Alternatively, interpreting the Economic Dualism's undervalued other half, emerges through an analysis of the institutions of social reproduction. Social reproduction refers both to the biological reproduction of the species and the reproduction of labour power. In any given society, it merges in relation to the institutions of childcare, eldercare, health care and education (Silvey 2009). Some feminists have even suggested 'that social reproduction is integral to life itself. It is the "fleshy, messy, and indeterminate stuff of everyday life"' (Katz 2001b, p. 711, cited in Silvey 2009, p. 510). Mellor (1997) conceives of it as care for the sick, the needy, the old, and the young, where emotions, feelings and wisdom are valued equally.

However, the economy, according to Mellor (1997), only wants 'man' when he is fit. It does not want his sickness, or childhood, or need for rest or sleep, or hunger, or worries, or dirty clothes, or parenting, his responsibilities or his

Economic Dualism

Economic 'man'	Women's work
Market value	Subsistence
Personal wealth	Social reciprocity
Able-bodied workers	Sick, needy, old, young
Labour, intellect	Body (life of the body)
Exploitable resources	Ecosystems, wild nature
Tradeable knowledge	Feelings, emotions, wisdom

Figure 5.3 Economic Dualism
Source: Mellor 2012 (used with permission).

ageing. 'Man here, of course, can be a woman. The gendering is in the construction of economic "man", it is not an absolute biological divide' (Mellor 1997, p. 135). What the construction of 'economic man' denies is 'the embodiedness of human beings in their biological functioning and the embeddedness of humanity in its ecological context' (Mellor 1997, p. 130). There is a point of difference that exists within embodiment, and that is that men are subject to the same realities as women, but women bear the burden of that embodiment (Mellor 1997).

This first sub-section has provided an overview of ecofeminism's historical origins and dismissed calls of its day as passing as premature. A retrospective analysis on these alternative models of 'economy' do not fall into the realm of myth or fable as arch rival neoliberal economics has demonstrated. Rather these representations offer invaluable tools to analyse the deeper gendered political economy of REDD, where women in developing economies have been invisibilised as silent offsets under this, so far, incentives-based scheme. For those who are incapable of moving on from the essentialism question, Mellor's (1997) suggests that 'an essentialist debate about ultimate differences between men and women is more helpfully redrawn in terms of materialist analysis of humanity's engagement with its own temporality rather than its particular form' (Mellor 1997, p. 136).

Converging theoretically: UN-REDD, World Bank and what this means for women in REDD+ countries

It is worth recalling the origins of the normative bases in the international system and the different post-World War II imperatives in their relationship with REDD+ and women. Particularly in light of when the World Bank announced at the UNFCCC COP 13 in Bali 2007 that it was seeking to work with the UN-REDD Programme as well as in 'harmonising normative frameworks' (UN-REDD n.d., cited in World Rainforest Movement 2010). The difference, according to Bessis (2003), results in the World Bank and the United Nations elaborating 'their policies on women's rights by means of entirely different theoretical approaches to the notion of gender' (Bessis 2003, p. 635).

88 'Silent offsets' and feminist perspectives

The international system, as understood within the context of organisational discourses and strategies, is partially homogenous (Bessis 2003). That is, the United Nations and its agencies, and the Bretton Woods system (whose organisations include the World Bank, the International Monetary Fund and today's World Trade Organization), are theoretically part of the same group of institutions. However, their post-war imperatives differed from the UN and this, according to Bessis (2003), had important implications for the way that women's issues were first approached by the World Bank, which was 'in a very different manner than the United Nations' (2003, p. 635).

On the one hand, economic and financial responsibilities assigned to institutions by the Bretton Woods 'founding fathers', were deliberately disconnected from the social sphere, in order that dominant interests be satisfied. The United Nations' mission, on the other hand, is to put into practice its founding act, the Universal Declaration of Human Rights (Bessis 2003). This involved putting an end to all forms of discrimination of human beings, regardless of ethnicity, gender, race and religion. Although the Declaration of Human Rights did not specifically mention women, 'Man' was supposed to cover both genders. Further, when it was finally realised that 'women were so invisible that they disappeared', and that women suffered discriminations over and above those suffered by men, that is when the Convention on the Elimination of All Forms of Discrimination Against Women was introduced and adopted (E. Pittaway 2016, pers. comm., 25 May).

It was not until 1975 that the UN started to take the 'woman question' seriously, with the first international conference devoted to the 'second sex' held in Mexico that year. Since then, global gender equality has been an official UN priority. Rather than apply an exhaustive analysis of the voluminous scholarly literature on the subject of gender policies at the international level, Bessis (2003) instead maps the general patterns of international thinking about feminism, women and gender over the subsequent two and a half decades. In summary, ever since the 'gender approach' has been an unavoidable aspect of development policy, the gender analyses that have emerged from the WB and the UN have been ideologically split. As a consequence of its different imperative heritage, the WB conceptualised women as a 'new type of economic actor', one potentially able to guarantee social stability in an era when stability is increasingly difficult to achieve. Women's rights were thus secondary. That is, women were instrumentalised by the WB in the sense that the promotion of women 'is not an end itself but rather a means of implementing the bank's policies for economic growth and the eradication of poverty' (Bessis 2003, p. 641).

However, the WB did not have a monopoly on instrumentalising women. Leach (2007) argues that the ecofeminists in their portrayal of women as close to nature served otherwise instrumental ends; that maintaining the idea that women are especially and inherently close to nature could only be sustained by the strategic interests it served. WED and their 'ecofeminist fables' (Leach 2007, p. 74) together supported a view that targeted women as prime movers in

'Silent offsets' and feminist perspectives 89

resource conservation projects. For example, a UNDP official commented on the justification for a project where women's labour promoted fruit tree agroforestry in the Gambia as '"women are the sole conservators of the land"' (Leach 2007, p. 72). At the same time, the WB had developed a 'win-win' approach to gender and the environment, arguing for a more general identity of interest between environmental resources and women, and 'thus treating women as the best agents for ensuring resources conservation' (Leach 2007, p. 72).

With the best intentions on the part of UN agencies, such as the UNDP translating ecofeminist ideas into development practice going awry, the global tendency of integrating women into development processes continues 'without attempting to overturn the logic that produced the inequalities in the first place' (Bessis 2003, p. 643). Further,

> even if gender discourse had earned the right to be taken into account, neither the UN agencies nor the WB has questioned fundamentally the androcentric logic of the bank's macroeconomic interventions schemes [structural adjustment programmes], its political tools, its work in the field, or its project infrastructure.
>
> (Bessis 2003, p. 643)

Quite some time has passed since Bessis (2003) undertook her sweeping review of international institutional thinking about women, gender and feminism since 1975. However, there is little evidence of this type of androcentric analysis being undertaken in relation to UN-REDD and the World Bank FCPF.

To minimise ecofeminism to its essentialist mode is also to overlook other important economic insights that address the deeply androcentric logic that the WB and the UN appear paralysed in adequately addressing. Two examples will suffice here. Gender equity is mentioned neither in the 1992 United Nations Framework Convention on Climate Change nor in the 1997 Kyoto Protocol (Gender CC 2011). With respect to gender relations in climate change negotiations, as of 2006, 'not the Parties, the Observers, nor the UNFCCC Secretariat has undertaken efforts to integrate gender aspects into the negotiations. The same is true for the Intergovernmental Panel on Climate Change' (Röhr 2006, p. 9). It has not been until recently, however, that some Parties from Annex II countries became aware of the necessity to include gender equality in the debates. A breakthrough was reached in Bali at COP 13 in 2007 (Hemmati & Röhr 2009).

Around the same time gender was absent through the UNFCCC process, the World Bank were not as reticent in defining 'gender equality as smart economics' in its 2006 *Gender Action Plan* (Rooke 2009, p. 1). In fact, the WB had not deviated from that view in over two decades, when the principal arguments presented in 1987 to justify their conversion to a gender approach were as economic and strategic (Bessis 2003) then. However, two short years after the publication of the 2006 *Gender Action Plan*, and the notion of gender

90 *'Silent offsets' and feminist perspectives*

equality disappears under the World Bank's Climate Investment Funds (CIFs) launched in 2008 (the Strategic Climate Fund and the Clean Technology Fund). These funds 'never mention gender or women in relation to the funds' objectives, governance, project criteria, evaluation measures or budget targets with one exception ... for selecting adaptation expert group members under the Pilot Program for Climate Resilience [under the Strategic Climate Fund]' (Marston 2009).

Based on this analysis of gender/women/feminist thinking in international institutions, it is apparent that unless and until the WB and the UN address the underlying androcentric logic inherent in both, as Bessis (2003) suggested well over a decade ago and the ecofeminists have been arguing since the mid-1970s, then women's rights issues will continue to be presented as 'the now familiar rhetoric on gender', thus risking further being an 'alibi for inaction' (Bessis 2003). Addressing the androcentric logic at the heart of the international institutions requires the conceptual intervention on Mellor's (2012) 'Economic Dualism'. What is unique about this alternative representation of 'economy' (provisioning), is that it captures the distinctiveness of women's work which 'is its existence in *time*' (Mellor 1997, p. 136, italics in original).

Orthodox economic models are not good at factoring in the concept of time. Nicholas Georgescu-Roegen demonstrated that they exclude the irreversibility of time (Latouche 2010). 'Economic models occur in mechanical and reversible time. They ignore entropy that is, the irreversibility of matter and energy conversions. By eliminating the earth from production functions, around 1880, the ultimate bond with nature was broken' (Latouche 2010, p. 520).[1] And, one might add, excluding social reproduction from production was where the ultimate bond with women was broken from Smith to Marx on.

Not only do current economic models ignore the irreversibility of time, they also dismiss 'biological time' in favour of 'clock time'. Mellor (1997) claimed that the most important element in the development of industrial time from that of 'free time' was the separation of biological time from clock time. That is, the time necessary to maintain human emotional and physiological existence, from 'clock time' (Mellor 1997). REDD+ projects, with their emphasis on payment for performance or 'results-based', demonstrated by emission reductions that are additional, are, in theory at least, time bound by baselines and time lines, and the like. However, REDD+ has no facility to factor in entropy, nor provide a pro-rata account of biological time in calculating how many tonnes of CO_2e (carbon dioxide equivalent) this may be worth. Calculating one's existence in time, the work of women is, accordingly, ignored.

The power of the Economic Dualism lies not in myths to sustain its dominance, but fact. Its notion of 'economy', one of provisioning, for the whole human community in an ecologically sustainable way, includes what the existing neoliberal economic models do not, thus sustaining one of global sustainability forest governance's enduring myths: 'what is counted – through valuation – counts' (Delabre et al. 2020, p. 5). As Mellor's (2012) dualism makes clear (Figure 5.3), the left side of the ledger, 'economic man', cannot be

'Silent offsets' and feminist perspectives   91

sustained without the right side, women's work. Her labour and her existence, under market-based solutions to climate change, such as REDD+, is accordingly treated as an externality in the silent offset economy of carbon commodification. Market-based solutions, have locked-in a crediting mechanism like REDD+, at the foreclosure of alternative notions of sustainable ecological economy embedded in these metaphors. One can only imagine if ecofeminism had flourished 40 years ago and neoliberalism had withered on the vine, what REDD+ would be like today. To be sure, the notions of 'gender-neutral beings', 'sexually neutral equality beings', indeed 'silent offsets' would be obsolete, neoliberal nomenclature.

(2) Operationalising REDD+ – the practical impacts on women of the South and their construction as 'silent offsets'

As section 1 has presented, the ideological construction of women of the global South has been conceptually and materially structured as silent offsets in REDD+. Indeed, this gendered analysis can apply more broadly to climate change mitigation crediting mechanisms involving offsets in general. This author draws on her own *in situ* research conducted in Indonesia in 2014–2015 entitled 'The impacts of the UN-REDD+ Program on Indigenous Women in Indonesia', to provide some material insight, thus opening one of the book's major hinges: moving the discussion from 'conceptual research' (M Wearing 2016, pers. comm., 25 May) into practice.

What section 1 has shown us is that the notion of 'silent offsets' has seen no singular linear theoretical pathway to its definition. What has emerged, thereby serving as an additional meaning to what women as 'silent offsets' means, is that forest-dependent women and indigenous communities of the global South whose countries are participating partners in REDD+, are the prime subjects; embedded – but not embodied – in a global economy that privileges the reproduction of carbon capital, when 'what matters is to move towards an economy that promotes a broader [social] reproduction of life' (Latin American Network of Women Transforming the Economy, cited in Filippini 2010). Mellor's (2012) Economic Dualism and other metaphors for 'economy' are powerful visual reminders of the foundational falsity upon which monetary economies are built – tips and icing, where women and the natural world represent the subordinated half of the Economic Dualism (Mellor 1997, 2012; Figure 5.3), the bottom two layers of Henderson's (1982) Layer Cake with Icing (Figure 5.1), and the submerged Iceberg Model (Figure 5.2).

For better or for worse, dualism remains the major conceptual division (Cameron & Gibson-Graham 2003) central to all three models. However, translating what this means for women in REDD+ programmes, pilot and readiness phase projects in developing countries requires 'adding on' another layer (Cameron & Gibson-Graham 2003), thus rendering what is invisible, or the 'silent' component of what is a silent offset, visible in the hierarchical depiction of the industrial productive economy.

92 'Silent offsets' and feminist perspectives

Drawing on theoretical insights emerging from section 1, this second and final section offers up a critique of an early publication commissioned by the UN-REDD Programme in 2011, under the guidance of the UNDP's UN-REDD and Gender Teams, entitled 'The Business Case for Mainstreaming Gender in REDD+' (December 2011). The framework for REDD+ had been agreed upon the previous year at the UN climate talks in 2010. However, the business case its authors adopt, legitimises, inadvertently or otherwise, the construction of indigenous communities and women in participating countries host to REDD+ projects as part of a gendered silent offset economy.

An analysis of gender in operationalising UN-REDD+ policies

At the COP 16 in Cancun in 2010, the framework for the REDD+ mechanism was agreed upon by the Parties in what is referred to as the 'Cancun Agreements' (UN-REDD Programme 2011, p. 8). The REDD texts were the Outcome of the work of the Ad Hoc Working Group on Long-term Cooperative Action (AWG-LCA). Depending on national circumstances, REDD+ is to be phased in over three stages: (1) national strategies [or action plans], policies and capacity building; (2) implementation of strategies and policies of results-based demonstration activities; and (3) results-based actions that should be fully measured, reported and verified (MRV) (Lang 2012b). Additionally, the national strategy should address forest degradation, drivers of deforestation, forest governance issues, land tenure issues, gender considerations and safeguards (Lang 2012b).

In operationalising the REDD+ phases, UN organisations, including the FAO, UNDP and UNEP working in collaboration under the UN-REDD Programme, are assisting national governments in their preparation of REDD+ programmes which imply complex strategies of MRV (Nunez 2011). '[T]he World Bank has been leading the provision of economic incentives through the Bank's FCPF' (Nunez 2011). Notwithstanding serious flaws identified in a report by the World Bank's Independent Evaluation Group (IEG) regarding the FCPF in August 2012,[2] five years after its launch at COP 13 in Bali in 2007 and a period in which the carbon market collapsed (Bond 2011), the flaws relate more particularly to what REDD+ already means in practice for women in participating countries. Most country members that the FCPF has supported through its REDD+ 'Readiness Fund' are at an early stage in their development of REDD+ strategies. That is, of a total of 36 member countries, 24 countries have been supported with readiness strategies, with the World Bank having helped Indonesia, Nepal, the Democratic Republic of the Congo, the Republic of the Congo and Ghana move into the implementation phase of REDD.[3]

Where do women feature in REDD+ readiness? Before that question can be answered, another more structurally relevant one emerges in relation to the role of women's exclusion from the forest sector more generally (The Forests Dialogue 2012). REDD+ is essentially a forest-based mitigation programme.

However, despite decades of evidence-based research into the importance of the role women play in the protection and management of forests and in their use, they are excluded for a multitude of reasons. Women Organizing for Change in Agriculture and Natural Resource Management (WOCAN) co-hosted a forests dialogue in Nepal on the exclusion and inclusion of women in the forestry sector in September 2012 (The Forests Dialogue 2012). In a broad scoping paper, the exclusion of women from the forestry sector was (re)examined for its probable causes at the local, professional, institutional and policy levels.

At a local level, women's forest work largely revolves around subsistence work. The activities of women are largely invisible for several reasons. For example, low levels of literacy, where 'men in Indonesia voiced their perception of women who could not read or write had no business being leaders of forestry groups' (The Forests Dialogue 2012, p. 3). Further, the lack of women in environmental projects in Cambodia is because, as one group of women expressed, 'no one had invited them to do so' (The Forests Dialogue 2012, p. 3). This exclusion is also evident in the formal sector as a lack of women professionals in forest departments and the forest industry (The Forests Dialogue 2012, p. 2). Finally, we see women's exclusion from forestry at the policy and institutional level (The Forests Dialogue 2012, p. 2). What is striking about the scoping paper's recommendations is that many of the recommendations have been around since 1995. For example, in promoting local-level inclusion of women, forest departments should ensure a 30–50 per cent quorum of women on forest management projects (Sarin 1995, cited in The Forests Dialogue 2012). The paper ends with a series of 'Potential Fracture Lines to be Explored'. One such line suggested: 'Why has research and knowledge on women's roles in forest management not been effective in bringing about their inclusion in the forestry sector?' (The Forests Dialogue 2012). As one academic and international development scholar noted, it is not only in the forestry sector; 'for me this comes from 40 years of working in the region [Asia-Pacific], and then 12 working in Africa asking the same question. Where are the women – in anything, in any development, in any major challenges, change – where are the women? So that's why I'm always looking' (E Pittaway 2015, pers. comm., 31 March).

The UN-REDD Programme/UNDP advance several arguments for the business case of mainstreaming gender in REDD+, suggesting that it may: increase efficiency (defined here as reducing transaction costs for REDD+ programmes); increase efficacy (defined here as reducing greenhouse gas emissions that stem from forest and land use); and increase sustainability (ensuring permanence of mitigation benefits, thereby reducing the risk of reversals for project investors, be they public or private) (UN-REDD Programme 2011, p. 9).

While the intention is not to minimise the goal of gender equality in climate change solutions, the arguments advanced by the authors of the business case for mainstreaming gender in REDD+ requires closer scrutiny for several reasons.

94 *'Silent offsets' and feminist perspectives*

With respect to the increased efficiency argument that gender mainstreaming REDD+ 'may' deliver, McAfee (2012) diminishes this argument on a more transactional level as it relates to the logic employed by institutional economics. That is, there is an inverse relationship between the size of the project and the transaction costs involved in the construction and maintenance of markets. For example, estimating and pricing ecosystem function costs, matching buyers and sellers, assessing compliance with conservation guidelines and ensuring payments are made. Further,

> [d]iseconomies of scale make it more expensive to enrol and monitor many smallholders than to pay fewer, larger-scale landholders. Transaction costs may be even higher when the intended recipients of PES lack individual property rights or abide by complex, overlapping systems of tenure or usufruct rules, as is common among forest-dwelling peoples.
>
> (McAfee 2012, p. 117)

With respect to women in REDD+ countries, the impact of market-based mechanisms is two-fold. Firstly, in monetary economies, depicted in the earlier metaphors, women generally occupy a marginal position due to their role as caretakers of families, in looking after the old, in charge of raising children, cooking and procuring water. Secondly, women seldom participate in the settlement of transactions (Nunez 2011).

A second argument for gender mainstreaming REDD+, according to the UNDP business case, relates to the efficacy or reduction of greenhouse gas emissions stemming from forest and land use. What is known already from the analysis of the CDM, which REDD+ is modelled on, should serve as a precautionary warning, indeed lesson, that offsetting emissions is a zero-sum game. That is, offsetting, by definition, is designed to produce a zero-sum outcome, not reducing emissions, but simply moving emissions where reductions are made (Gerebizza 2011). How mainstreaming gender in REDD+ can eliminate this inherent structural flaw of crediting mechanisms, indeed the whole counterfactual illusion central to offsetting, is not clear, and nor do the authors of the business case make this so.

The third argument deployed for gender mainstreaming REDD+ centres on a definition of increased 'sustainability' that is irreducibly circumscribed by neoliberal logics. This is most evident in the business case's circumscription of sustainability of mitigation benefits as, in the end, the preservation of investor confidence. In conclusion, the authors recommend:

> Overall, giving consideration to gender equality in each readiness component of REDD+ makes good business sense, both creating and benefiting from a more stable investment environment for forest carbon assets.
>
> (UN-REDD Programme 2011, p. 7)

Indeed, where are the women in REDD+?

'Silent offsets' and feminist perspectives 95

In summary, gender equality in REDD+ is ostensibly meant to be achieved through efficiency, efficacy and sustainability arguments that make its mainstreaming good business sense. Read in this light, the rhetoric on 'gender' does not merely risk further becoming an alibi for inaction as Bessis (2003) suggested, but is actively used to legitimise neo-imperial geographical hierarchies embedded in the 'economic man' half of the economic dualism. The other half of the dualism – the language of women's work, subsistence, social reciprocity, care of the sick, needy, old and young, body – the life of the body, ecosystems and wild nature, feelings, emotions and wisdom (Mellor 2012), may be intermittently referred to in the business case (UN-REDD Programme 2011). However, within the broader neoliberal economism in which the business case is nestled, this half externalises, oppresses, renders invisible and ultimately silences women. In the case of REDD+, women remain enclosed in a conceptual silo that privileges carbon capital production; not social reproduction that, if the authors chose, they too could have made 'The social reproduction case for mainstreaming gender in REDD+'.

Similarly, the Economic Dualism model presents an inherent bias in favour of its prioritisation of 'economic man'. The critique of such dualisms has mainly come from feminists who argue that such a manner of conceptualising neither represents a simple discrete statement of difference (A, B, …), nor is constructed on the basis of an analysis of 'the interrelations between the objects being defined (capital: labour). 'It is a dichotomy specified in terms of a presence or an absence; a dualism which takes the classic form of A/not-A' (Massey 1992, p. 71). Only one of the terms (A) within this kind of conceptualisation is positively defined, whereas the other term (not-A) is perceived solely in relation to A, and as lacking in A (Massey 1992). When mapping Massey's A/not-A classic dualism (Massey 1992) onto Mellor's (2012) Economic Dualism, the theoretical and practical implications of this dualism on women in the global South, whose countries are participating in REDD+, are two-fold. Firstly, and theoretically, women of the South participating in REDD+ national programmes and projects (not-A), are conceived only in relation to A defined by, in this dualistic conception, 'economic man'. At every level of the Economic Dualism presented by Mellor (2012) – economic man, market value, personal wealth, able-bodied workers, labour, intellect, exploitable resources, tradeable knowledge (or A) – there is no relational analysis to 'not-A' because they are conceived as 'lacking' in A at all.

Secondly, the practical implications for REDD+ women caught on the 'wrong side' of the classic dualism's ideological ledger are marked by absence or exclusion in REDD+ practice; beyond tokenistic processes of 'participation', exclusions continue to occur (Delabre et al. 2020).

Conclusion

Defining what a silent offset means in the context of this critical appraisal of REDD+'s transactional and technical structure has introduced a

96 *'Silent offsets' and feminist perspectives*

reinterpretation of ideas and alternative theories such as ecofeminism, decades old, but nonetheless relevant to the one-sided notion of 'economy' under the dominant neoliberal economic orthodoxy. REDD+, shaped within the shadow of the existing dominant neoliberal model of economics, eschews the idea of Mellor's (2012) alternative notion of an ecofeminist provisioning economy. Further, this requires integrating both sides of the economic dualism, thus giving legitimacy to the REDD+ Programme as one which includes its co-benefits meaningfully: nature and social reproduction. In order to be legitimate in the eyes of those who are impacted by such projects, REDD+ requires further analysis beyond the theoretical notion of gender mainstreaming for women's inclusion that has failed to engage with indigenous communities and women who are its subjects.

Notes

1 According to ecological economist Herman Daly, over-consumption of natural capital will only cease when natural capital's definition is restored to its original meaning, as intended by the classical economists Adam Smith, David Ricardo, Thomas Malthus amongst others, at a time when the representation of the natural world was 'land' (Daly, cited in McDaniel 2005). In the 20th century, economists bestowed 'capital equivalence' on traditional capital – labour and capital, that is the structures and equipment employed to provide services and produce goods – with 'land' or natural capital, for example, soil fertility, forests, water, fisheries, capacity to assimilate waste. As a major tenet of neoliberal economics, natural capital and human-made capital were thus lumped together as one (Daly, cited in McDaniel 2005). However, decoupling that false conflation of natural capital and human-made capital, where the natural world is erroneously subsumed within the 'total capital stock', will require the more ambitious project of decoupling the neoliberal economic order's narrow view of civilisation through an economic prism (Ralston Saul 2006). 'The flawed logic of this capital equivalence permitted economists to assume falsely that the human economy subsumed the natural world merely as a substitutable factor in production. For the economist, the natural world became part of the total capital stock available for production and its "products and services", as priced in the market transactions became its value. Nature was reduced to a commodity' (Daly, cited in McDaniel 2005, p. 136).

2 "REDD+ is a more expensive, complex, and protracted undertaking than was anticipated at the time of the FCPF's launch," the IEG writes. What the IEG fails to point out is that this should come as no surprise. The Bank undertook no feasibility study, prepared no business case, and did no market or technical analysis before launching the FCPF in Bali [at COP 13 in 2007] ... The IEG recommends that the FCPF "needs to update and clarify its mission to the World Bank's Board and to its participating members in relation to the changes that are taking place in the carbon market" ... The IEG also recommends that the World Bank "needs a high-level strategic discussion on its overall approach to REDD" and points out that the Bank "faces a risk to its reputation in case financing does not materialise on the scale envisaged". A final recommendation is that the Bank should focus on activities such as "legal and policy support for land tenure and forest governance reforms"' (IEG 2012, cited in Lang 2012a).

3 It is worth noting that 'the FCPF has so far spent about US$22 million to deliver a total of US$4.9 million in grants. Of this, 70% went to five countries' (REDD Monitor 2012).

References

Barnett, J & O'Neill, S 2010, 'Maladaptation', *Global Environmental Change*, vol. 20, pp. 211–213.

Bennholdt-Thomsen, V & Mies, M 1999, *The subsistence perspective: beyond the globalised economy*, Zed Books, London UK & New York, USA.

Bessis, S 2003, 'International organisations and gender: new paradigms and old habits', *Signs: Journal of Women in Culture and Society*, vol. 29, no. 2, pp. 633–647.

Boas, H (ed) 2011, *No REDD papers: volume one*, e-book, accessed 15 June 2017, www.ienearth.org/docs/No-Redd-Papers.pdf.

Bond, P 2011, *Patrick Bond speaking at OccupyCOP17*, 10 December 2011, online video, accessed 12 December 2011, https://youtu.be/gX2Pn6b_YuQ.

Bumpus, A & Liverman, D 2008, 'Accumulation by decarbonisation and the governance of carbon offsets', *Economic Geography*, vol. 84, no. 2, pp. 127–155.

Burton, I 2009, 'Deconstructing adaptation ... and reconstructing', in ELF Schipper & I Burton (eds), *The Earthscan reader on climate change*, Earthscan, London, p. 11.

Cabello, J & Gilbertson, T (eds) 2010, *No REDD!: a reader*, e-book, accessed 15 June 2017, http://no-redd.com/wp-content/uploads/2015/01/REDDreaderEN.pdf.

Cameron, J & Gibson-Graham, J 2003, 'Feminising the economy: metaphors, strategies, politics', *Gender, Place and Culture: A Journal of Feminist Geography*, vol. 10, no. 2, pp. 145–157.

Carvajal-Escobar, Y, Quintero-Angel, M & Garcia-Vargas, M 2008, 'Women's role in adapting to climate change and variability', *Advances in Geosciences*, vol. 14, pp. 277–280.

Daly, H 1995, 'The irrationality of homo economicus', *Developing Ideas Digest*, Developing Ideas interview by Karl Hansen, College Park, Maryland, USA, 8 February 1995, accessed 3 June 2017, http://pratclif.com/sustainability/Herman%20Daly.htm.

Daly, H 2005, 'Living in a finite world: Herman Daly and economics', in CN McDaniel, *Wisdom for a livable planet: the visionary work of Terri Swearingen, Dave Foreman, Wes Jackson, Helena Norberg-Hodge, Werner Fornos, Herman Daly, Stephen Schneider, and David Orr*, Trinity University Press, San Antonio.

Delabre, I, Boyd, E, Brockhaus, M, Carton, W, Krause, T, Newell, P, Wong, GY & Zelli, F 2020, 'Unearthing the myths of global sustainable forest governance', *Global Sustainability*, vol. 3, no. 16, pp. 1–10.

Dufrasne, G 2018, 'COP24: Withdrawn UN advert shows why carbon offset scheme should be scrapped', *Carbon Market Watch*, accessed 19 August 2020, https://carbonmarketwatch.org/2018/08/31/withdrawn-un-advert-shows-why-carbon-offset-scheme-should-be-scrapped/.

Filippini, A 2010, 'Women and climate change in Cochabamba', *World Rainforest Movement*, no. 154, accessed 10 June 2017, http://wrm.org.uy/bulletins/issue-154/.

Filippini, A 2011, 'Women and climate change in Cochabamba', *Climate Justice Now!*, accessed 13 May 2017, www.climatejusticenow.org/women-and-climate-change-in-cochabamba/.

Fraser, N 2008, 'Reframing justice in a globalizing world', in N Fraser, *Scales of justice: reimagining political space in a globalizing world*, Polity Press, Cambridge, UK.

Fraser, N 2020, 'Taking care of each other is essential work', Nancy Fraser interview by *VICE*, Clio Chang, 7 April 2020, accessed 1 May 2020, vice.com.

Gender CC 2011, 'Climate Finance', *Gender CC: Women for Climate Change*, accessed 10 June 2017, http://gendercc.net/genderunfccc/topics/climate-finance.html.

98 'Silent offsets' and feminist perspectives

Gerebizza, E 2011, 'Can financial markets solve the climate market? unpacking the false solutions by the World Bank', *REDD-Monitor*, accessed 17 June 2017, www.redd-m onitor.org/wp-content/uploads/2011/03/Discussion_note-march2011.pdf.

Gilbertson, T 2010, 'Fast forest cash: how REDD+ will be market-based', in J Cabello & T Gilbertson (eds), *No REDD!: a reader*, e-book, accessed 15 June 2017, http:// wrm.org.uy/wp-content/uploads/2013/04/REDDreaderEN.pdf.

Gioli, G & Milan, A 2018, 'Gender, migration and (global) environmental change', in R McLeman & F Gemenne (eds), *Routledge Handbook of Environmental Displacement and Migration*, Routledge, London.

Goldtooth, TBK 2009, 'The REDD train is going pretty fast and it's left us at the station', accessed 8 June 2017, www.redd-monitor.org/2009/01/14/interview-with-tom -bk-goldtooth/.

Goldtooth, T 2010, 'Why REDD/REDD+ is not a solution', in J Cabello & T Gilbertson (eds), *No REDD!: a reader*, e-book, accessed 15 June 2017, http://wrm.org.uy/wp -content/uploads/2013/04/REDDreaderEN.pdf.

Hall, S 2011, 'The neoliberal revolution: Thatcher, Blair, Cameron – the long march of neoliberalism continues', *Soundings: A journal of politics and culture*, no. 48, Summer, pp. 9–27.

Harvey, D 2011, *Spatial justice workshop – David Harvey*, online video, University of Westminster, accessed 3 June 2012, www.youtube.com/watch?v=3WpR4exvm8g.

Hemmati, M & Röhr, U 2009, 'Engendering the climate-change negotiations: experiences, challenges, and steps forward', *Gender & Development*, vol. 17, no. 1, pp. 19–32.

Henderson, H 1982, *Total productive system of an industrial society: layer cake with icing*, digital image, accessed 14 May 2017, http://hazelhenderson.com/wp-content/up loads/totalProductiveSystemIndustrialSociety.jpg

Jackson, C 1993, 'Doing what comes naturally? women and environment in development', *World Development*, vol. 21, no. 12, pp. 1947–1963.

Jackson, C 1995, 'Radical environmental myths: a gender perspective', *New Left Review*, no. 210, Mar–Apr, pp. 124–140.

Karsenty, A & Ongolo, S 2011, 'Can "fragile states" decide to reduce their deforestation? The inappropriate use of the theory of incentives with respect to the REDD mechanism', *Forest Policy and Economics*, vol. 18, accessed 30 May 2011, doi:10.1016/j.forpol.2011.05.006, pp. 38–45.

Katz, C 2001, 'On the grounds of globalization: a topography for feminist political engagement', *Signs: Journal of Women in Culture and Society*, vol. 26, no. 4, pp. 1213–1234.

Kuletz, V 1992, 'Eco-feminist philosophy interview with Barbara Holland-Cunz', *Capitalism Nature Socialism*, vol. 3, no. 2, pp. 63–78.

Lang, C 2009, 'Forests, carbon markets and hot air: why the carbon in forests should not be traded', in S Bohm & S Siddhartha (eds), *Upsetting the offset: the political economy of carbon markets*, e-book, accessed 21 May 2017, http://mayflybooks.org/ wp-content/uploads/2010/07/9781906948078UpsettingtheOffset.pdf.

Lang, C 2012a, 'Down the rabbit hole: REDD in the CDM', *REDD-Monitor*, weblog post, 12 September, accessed 10 May 2017, www.redd-monitor.org/2012/09/12/ down-the-rabbit-hole-redd-in-the-cdm/#more-12907.

Lang, C 2012b, 'A short guide to REDD in the UN climate negotiations ahead of Doha', weblog post, 8 November, accessed 10 May 2012, www.redd-monitor.org/ 2012/11/08/a-short-guide-to-redd-in-the-un-climate-negotiations-ahead-of-doha/?utm_ source=feedburner&utm_medium=email&utm_campaign=Feed%3A+Redd-monitor+%28REDD-Monitor%29.

Latouche, S 2010, 'Degrowth', *Journal of Cleaner Production*, vol. 18, no. 6, pp. 519–522.

Leach, M 2007, 'Earth mother myths and other ecofeminist fables: how a strategic notion rose and fell', *Development and Change*, vol. 38, no. 1, pp. 67–85.

Lohmann, L 2011a, 'Capital and climate change', *Development and Change*, vol. 42, no. 2, pp. 649–668.

Lohmann, L 2011b, 'The endless algebra of climate change markets', *Capitalism Nature Socialism*, vol. 22, no. 4, pp. 93–116.

Lovera, S 2009, 'REDD realities', in U Brand, N Bullard, E Lander & T Mueller (eds), *Contours of climate justice: ideas for shaping new climate and energy policy*, Critical Currents, no. 6, pp. 46–53.

MacGregor, S 2009, 'A stranger silence still: the need for feminist social research on climate change', *The Sociological Review*, vol. 57, no. 2s, pp. 124–140.

MacGregor, S 2014, *Heat and light: why environmentalism (still) needs feminism*, Lunchtime Colloquium, accessed 17 May 2020, www.carsoncenter.unimuenchen.de/events_conf_seminars/event_history/2014-events/2014_lc/index.html.

Marston, A 2009, *Will rights and gender be at the heart of World Bank's climate response*, Update 66, Bretton Woods Project, accessed 27 May 2017, http://old.brettonwoodsproject.org/art-564819.

Massey, D 1992, 'Politics and space/time', *New Left Review*, no. 196, Nov/Dec, pp. 65–84.

Massey, D 2011, *Spatial justice workshop – Doreen Massey*, online video, University of Westminster, accessed 3 June 2012, www.youtube.com/watch?v=l7vppCnTu88.

Maxton-Lee, B 2018, 'Narratives of sustainability: a lesson from Indonesia: global institutions are seeking to shape an understanding of sustainability that undermines its challenge to their world view', *Soundings: A journal of politics and culture*, no. 70, Winter, pp. 45–57.

Maxton-Lee, B 2020, 'Forests, carbon markets, and capitalism: how deforestation in Indonesia became a geo-political hornet's nest', Guest Post, *REDD-Monitor*, accessed 22 August 2020, https://redd-monitor.org/2020/08/21/guest-post-forests-carbon-markets-and-capitalism-how-deforestation-in-indonesia-became-a-geo-political-hornets-nest/.

McAfee, K 1999, 'Selling nature to save it? biodiversity and green developmentalism', *Environment and Planning D: Society and Space*, vol. 17, pp. 133–154.

McAfee, K 2012, 'The contradictory logic of global ecosystem services markets', *Development and Change*, vol. 43, no. 1, pp. 105–131.

McAfee, K 2014, *Green economy or buen vivir: can capitalism save itself*, Lunchtime Colloquium, Rachel Carson Centre for Environment and Society, online video, accessed 17 May 2020, www.carsoncenter.unimuenchen.de/events_conf_seminars/event_history/2014-events/2014_lc/index.html.

McDaniel, C 2005, *Wisdom for a livable planet: the visionary work of Terri Swearingen, Dave Foreman, Wes Jackson, Helena Norberg-Hodge, Werner Fornos, Herman Daly, Stephen Schneider, and David Orr*, Trinity University Press, San Antonio.

Mellor, M 1992, 'Eco-feminism and eco-socialism: dilemmas of essentialism and materialism', *Capitalism Nature Socialism*, vol. 3, no. 2, pp. 43–62.

Mellor, M 1996, 'Myths and realities: a reply to Cecile Jackson', *New Left Review*, no. 217, May/Jun, pp. 132–137.

Mellor, M 1997, 'Women, nature and the social construction of "economic man"', *Ecological Economics*, vol. 20, no. 2, pp. 129–140.

100 *'Silent offsets' and feminist perspectives*

Mellor, M 2012, *Just banking conference*, Friends of the Earth Scotland, online video, accessed 14 May 2017, www.youtube.com/watch?v=i-9vMXfTM10.

Mies, M 2007, 'Patriarchy and accumulation on a world scale revisited: keynote lecture at the Green Economics Institute, Reading, 29 October 2005', *International Journal of Green Economics*, vol. 1, no. 3–4, pp. 268–275.

Mousseau, F 2019, *The Highest Bidder Takes It All: The World Bank's Scheme to Privatize the Commons*, Oakland Institute, accessed 21 August 2020, www.oaklandin stitute.org/highest-bidder-takes-all-world-banks-scheme-privatize-commons.

Mustalahti, I, Bolin, A, Boyd, E & Paavola, J 2012, 'Can REDD+ reconcile local priorities and needs with global mitigation benefits? lessons from Angai Forest, Tanzania', *Ecology and Society*, vol. 17, no. 1, pp. 114–125.

Nunez, R 2011, 'The sins of the REDD+ approach', *World Rainforest Movement*, no. 169, accessed 10 June 2017, http://wrm.org.uy/wp-content/uploads/2011/08/Bulle tin169.pdf.

Phelps, J, Guerrero, M, Dalabajan, D, Young, B & Webb, E 2010, 'What makes a "REDD" country?', *Global Environmental Change*, vol. 20, no. 2, pp. 322–332.

Pielke, R 2009, 'Rethinking the role of adaptation in climate policy', in ELF Schipper & I Burton (eds), *The Earthscan reader on climate change*, Earthscan, London, p. 345.

Ralston Saul, J 2006, *The collapse of globalism: and the reinvention of the world*, Penguin Books, Camberwell, Victoria.

REDD Monitor 2012, 'Independent Evaluation Group review of the FCPF: "World Bank needs a high-level strategic discussion on its overall approach to REDD"', 22 November, https://redd-monitor.org/2012/11/22/independent-evaluation-group-revie w-of-the-fcpf-world-bank-needs-a-high-level-strategic-discussion-on-its-overall-appro ach-to-redd/.

Redman, J 2008, *World Bank: climate profiteer*, Sustainable Energy and Economy Network, Institute for Policy Studies, accessed 21 May 2017, www.ips-dc.org/wp -content/uploads/2008/04/1207774611WPCP-SEEN-April-9-08.pdf.

Röhr, U 2006, 'Gender relations in international climate change negotiations', *Berlin: LIFE eV/genanet*.

Rooke, A 2009, 'Doubling the damage: World Bank climate investment funds undermine climate and gender justice', *Gender Action*, accessed 10 May 2017, www.gen deraction.org/images/2009.02_Doubling%20Damage_AR.pdf.

Sandel, M 2013, *What money can't buy: the moral limits of markets*, Farrar, Straus & Giroux, New York.

Schalatek, L, Bird, N & Brown, J 2010, 'Where's the money? the status of climate finance post-Copenhagen', *Climate Finance Policy Brief* No.1, accessed 6 June 2017, www.odi.org/sites/odi.org.uk/files/odi-assets/publications-opinion-files/5844.pdf.

Silvey, R 2009, 'Development and geography: anxious times, anaemic geographies, and migration', *Progress in Human Geography*, vol. 33, no. 4, pp. 507–515.

Stern, N 2006, *Stern review: the economics of climate change, summary of conclusions*, HM Treasury, UK Government Web Archive, accessed 10 May 2017, http://weba rchive.nationalarchives.gov.uk/20130129110402/http://www.hm-treasury.gov.uk/d/CL OSED_SHORT_executive_summary.pdf.

Stern, N 2009, *The global deal: climate change and the creation of a new era of progress and prosperity*, Public Affairs, New York.

The Forests Dialogue 2012, *Background paper for the scoping dialogue on the exclusion and inclusion of women in the forestry sector*, Women Organising for Climate Change in Agriculture and NRM, accessed 10 June 2017, http://theforestsdialogue.

org/publication/background-paper-scoping-dialogue-exclusion-and-inclusion-women-forestry-sector.

United Nations Programme on Reducing Emissions from Deforestation and Forest Degradation (UN-REDD Programme) 2011, *The business case for mainstreaming gender in REDD+*, The United Nations Collaborative Program on Reducing Emissions from Deforestation and Forest Degradation in Developing Countries, accessed 15 May 2017, www.undp.org/content/dam/undp/library/gender/Gender%20and%20Environment/Low_Res_Bus_Case_Mainstreaming%20Gender_REDD+.pdf.

World Rainforest Movement 2008, *From REDD to HEDD: WRM contribution to the Convention on Climate Change*, iss. 134, accessed 10 June 2017, http://wrm.org.uy/wp-content/uploads/2013/01/From_REDD_to_HEDD.pdf.

World Rainforest Movement 2010, *REDD in the Congo Basin*, December 2010, accessed 10 May 2017, http://wrm.org.uy/wp-content/uploads/2013/01/REDD_Congo.pdf.

Zuckerman, E 2007, *Huge gaps in the World Bank's gender action plan*, Bretton Woods Project, accessed 27 May 2017, www.brettonwoodsproject.org/2007/01/art-549094/.

6 Findings of the Indonesian study

Introduction

Introduced here are the general findings of a small-scale empirical study undertaken across several provincial capitals in Indonesia during the months of May, June and July 2014, entitled 'UN-REDD+ impacts on Indigenous women in Indonesia', referred to as the Indonesian study. An additional field visit to a 'closed' United Nations Programme on Reducing Emissions from Deforestation and Forest Degradation (UN-REDD+) project site in Central Kalimantan took place in February 2015.

The previous chapter presented a theoretically complex and abstract account of women as 'silent offsets' under the global mitigation programme of REDD+. By this account, social reproduction is leveraged to further the strategic interests of carbon capitalism, thus giving rise to the notion of a gendered silent offset economy. Forest-dependent women participating in REDD+ programmes, based on decades of research into the importance they play in the protection and management of forests for use (The Forests Dialogue 2012) and, in their allocated social roles, bear disproportionate impacts not only from climate change itself, but from programmes such as REDD+ that seek to address the problem of climate change, thus reinforcing the conditions that have produced it in the first place (Maxton-Lee 2018).

What became evident during the interviews and subsequent analysis of participant responses from the Indonesian study, however, is that women-specific findings were obscured by the more general findings presented here, though related. Accordingly, the following chapter will focus on the women-specific findings of the study.

Section 1 begins by providing an overview of the rationale for the study, its aims and research methodology. The Indonesian study's methodology of *in situ* research has, in a participatory research fashion, invited the voices of women and men experts in the REDD+ programme into this book. Section 2 presents the study's general outcomes.

Finally, a note on terminology. The preferred term for indigenous peoples in Indonesia is either customary communities or, in Bahasa Indonesian, masyarakat adat. Therefore, I use this term adat in respect of that preference. Perempuan means woman.

Findings of the Indonesian study 103

(1) Rationale, aims and research methodology of the Indonesian study

Rationale of the study

The study was undertaken to generate raw data to assess the impacts of the UN-REDD+ Programme activities in general, and to understand how adat women (perempuan adat) in particular are impacted. The data were gathered through semi-structured interviews conducted with 15 key informant women and men representative of customary communities involved in on-the-ground REDD+ readiness activities in Central Sulawesi, and a fully implemented UN-REDD+ demonstration activity pilot project in Central Kalimantan, Indonesia.

Desk-based insights and the empirical findings from this study have found that it is the solutions to climate change, in this study REDD+, and not climate change itself, that is leading to maladaptation and potentially displacement of peoples in participating REDD+ countries, such as Indonesia. Work et al.'s (2019) empirical research into both mitigation and adaptation policies and projects of forest conservation, irrigation and reforestation in Cambodia, while not focusing on displacement/migration as a result of maladaptation, found these climate change mitigation and adaptation (CCMA) projects and policies exhibited 'maladaptation in action', concluding with the claim – this book conceptually and empirically supports – that '[w]e face today new dangers from climate change projects and policies as much as we do from the effects of climate change itself' (Work et al. 2019, s59).

Information from case-based research has found that most people displaced from climate change will be within stated borders. That is, internally displaced and not international migrants (Oliver-Smith & Shen 2009). While the current literature addresses the gap on recognition for environmentally induced migration, there is a dearth of research on displacement whose main driver is the result of maladaptive climate change solutions, in particular *mitigation*.

Aims of the project:

1   To assess, through key informant interviews, whether maladaptation is present in on-the-ground REDD+ readiness activities and implemented programmes, and to establish if there is a gendered pattern.
2   To assess whether there is any evidence of migration/displacement resulting from REDD+ and to disaggregate this by gender.

Research methodology – feminist qualitative research

The Indonesian study employed qualitative research methods using semi-structured interviews with key informants involved in REDD+ in Indonesia. A sample size of 15 participants was designed to capture the expert knowledge of key informants involved in REDD+ and/or representatives of customary communities across three provincial capitals: Jakarta in Java; Palangkaraya in Central Kalimantan (Indonesian Borneo); and Palu in Central Sulawesi. The

104 *Findings of the Indonesian study*

latter two provinces were selected based on existing REDD+ pilot and readiness activities respectively, while Jakarta was selected on the basis of organisational affiliations with these two provinces as well as serving as the headquarters for some of the provincial organisations participating. Six participants were interviewed in Jakarta, four in Palangkaraya and five participants in Palu.

The range of participants occupied mainly middle to senior positions within their organisations with the exception of one independent researcher/adviser. All respondents were either formally or informally associated with non-profit organisational member groups such as Aliansi Masyarakat Adat Nusantara (AMAN) (Indigenous Peoples Alliance of the Archipelago), Perempuan AMAN (AMAN Women), Solidaritas Perempuan, HuMa (whose work focuses on the issue of law reform in the natural resources sector and was the research partner for the Indonesian study), Bank Information Centre, debtWatch, Bantaya, Libu Perempuan, AMAN Central Kalimantan, Lembaga Dayak Panarung and an independent researcher and adviser of Dayak communities. The latter participant was involved with UN-REDD+ activities in the joint Australian and Indonesian governments' Kalimantan Forests and Climate Partnership (KFCP) project from when on-the-ground activities were implemented in 2010 until the project's closure in June 2014.

These organisations were identified primarily on the basis of their alliance with, or representation of, customary communities involved in REDD+. In most cases, organisational involvement of REDD+ was not the core mission/activity of these various organisations and associations, but one of a range of activities involving capacity building and development, policy advocacy, education, campaigning, peace building, conflict resolution as well as other activities aligned with the social movement of masyarakat adat in Indonesia. With respect to their respective organisational involvement with REDD+, this varied from *in situ* research at REDD+ sites, to work around issues involving adat women and the impacts of climate change on them, monitoring projects and their implementation, programmes and policies of the World Bank on customary communities, such as the World Bank's convening/administrative role in the Forest Carbon Partnership Facility (FCPF), as well as tasks that involved translating project documents into popular language understood by masyarakat adat and education on globalisation more specifically. The range of activities listed here is not exhaustive, but broadly prescriptive of the work undertaken by a busy and dedicated network of non-governmental organisations (NGOs), associations and their affiliated members in a social movement dedicated to improving the well-being, safety, empowerment and dignity of customary communities in Indonesia.

To ensure the anonymity of the participants was preserved, their names, organisations and specific locality within the province, with the exception of providing the name of the provincial location (Central Kalimantan and Central Sulawesi) or capital (Jakarta), will not be identified in any of the direct quotations. The reason for including the name of the province is to provide the reader with some context of which stage the UN-REDD+ programme activities operate at. The de-identification process necessary for ethics approval of this study mandated that participant names, their organisations, position within the organisation, gender and specific locale not be identified. Initials of the participants and their province will be the only attributable reference used in their direct quotes or summaries of their responses, for example (RD Central Kalimantan 2014).

Originally, the study's design envisaged all-women respondents in an attempt to gain an all-women perspective, to gain a practical insight into what has been a deeply complex theoretical analysis of the notion of women as 'silent offsets', and who may be able to shed light on how 'gender bias' has emerged and is maintained in REDD+. However, upon arrival in Jakarta in May 2014, and in consultation with research partner HuMa, this all-female participant limitation proved difficult to implement on the ground in some of the provinces due to the lack of key informants with expert knowledge and involvement with the REDD+ programme. Notwithstanding this, of the 15 participants interviewed, 11 were women. The average duration of the interview was approximately one hour. With several of the participants, follow-up interviews of shorter duration were conducted. All interviews were tape recorded and later transcribed, in-country.

(2) Outcomes of the Indonesian study

The outcomes of the study as they relate to its aims are mixed and incomplete. Before elaborating on these specifics, it is necessary to recall how migration/ displacement due to maladaptation has been defined in this book: as a spatio-temporal manifestation of international mitigation and adaptation policies framed within a global carbon market framework, that are on a maladaptive trajectory. The temporal dimension of this driver is crucial, insofar as today's mitigation may be tomorrow's maladaptation. While experiences of climate change adaptation are expanding at pace, '[w]ithin this rapidly increasing research field, empirical studies highlighting maladaptation have emerged in recent years' (Juhola et al. 2016, pp. 135–136) (also see Magnan et al. 2016; Work et al. 2019). In other words, the consequences of the UN-REDD+ Programme on the potential migration pattern has not been fully realised – yet. Hence, the study's outcomes have raised further questions to be explored and addressed.

Summary of outcomes of the Indonesian study

This summary presents a short overview of the study's major themes, insights, findings, more generally described as outcomes, in relation to the Indonesian's study's two broad aims. The analyses of these outcomes are subsequently discussed.

106 *Findings of the Indonesian study*

Study Aim 1: To assess, through key informant interviews, whether maladaptation is present in on-the-ground REDD+ readiness activities and implemented programmes, and to establish if there is a gendered pattern.

Outcome: Based on the responses from all participants across the three sites, there is evidence to suggest that maladaptation is present in on-the-ground REDD+ demonstration and readiness activities. With respect to a pattern of gendered maladaptation, taken to mean a bias against women, the answer too is yes and this, as previously stated, will be addressed in the second findings chapter – 'Where are the women?'

Study Aim 2: To assess whether there is any evidence of *displacement/migration* resulting from REDD+ and to disaggregate this by gender.

Outcome: With respect to displacement/migration resulting from REDD+, these findings are mixed, depending on the provincial location where the study was undertaken and the stage at which REDD+ was operational.

Central Kalimantan: There was no evidence of migration as a result of the REDD+ pilot project under investigation in Central Kalimantan known as the KFCP. With respect to displacement of adat communities in terms of livelihood access, the results are more ambiguous for reasons explained shortly.

Central Sulawesi: The results on displacement/migration emerging from the study in Central Sulawesi are incomplete. That is because the REDD+ programme in Central Sulawesi was at the 'readiness' phase. In other words, no on-the-ground demonstration activities had been implemented at the time the investigation was undertaken in June 2014, although movement into this phase was imminent with the near-completion of REDD+ readiness activities. Hence, it is not yet possible to determine whether displacement/migration will occur as a result of REDD+ demonstration activities and is an area that is deserving of further research.

As a proxy, however, one scenario in terms of displacement of adat communities was raised by several of the Central Sulawesi respondents. That is, these responses centred on the potential extension of Lore Lindu National Park's boundaries to accommodate the potential expansion of the REDD+ project area. One village participating in the Free Prior and Informed Consent (FPIC) readiness activity is located in the sub-district of Gumbasa which shares a border with the national park (SB Central Sulawesi 2014). However, at the time of this study, no decision has been formally made as to the expansion of Lore Lindu's boundaries. I discuss this further in relation to the historical displacement that has occurred previously in this region (S Jakarta 2014; AP Jakarta 2014).

As a final note to the outcomes of the study's second aim, an often repeated narrative in participants' responses was the emphasis and relevance they apply to contextualising their responses against an historical backdrop. This is not only in relation to site-specific areas, where REDD+ readiness and demonstration activities are situated, but more broadly of Indonesia's political and legislative history that gave rise to the issues of deforestation and forest degradation, in which projects such as REDD+ are proposed and operate.

Findings of the Indonesian study 107

For many of the interviewees, it was implausible to respond to the semi-structured questions without first providing historical context, or some such other significant site-specific history, often at length. One respondent captured this trend well in asserting that one 'can't dissociate the historical facts with what's going on [with REDD+] today' (D Jakarta 2014).

Outcomes and analysis – Central Kalimantan, KFCP

In the province of Central Kalimantan, the REDD+ pilot project this study has focused on is the KFCP, located in the Kapuas district (district = kabupaten), east of the capital Palangkaraya. The project was conceived by the former Australian Federal Liberal National Party government, led by Prime Minister John Howard in 2007. The project was finally agreed to by the Federal Labor Kevin Rudd government and the President of the Republic of Indonesia on 13 June 2008 under the Indonesia-Australia Forest Carbon Partnership (IAFCP). Of a total funding pool of 40 million Australian dollars, the IAFCP allocated A $30 million to the KFCP. The objectives of the IAFCP were to formalise and build upon long-term practical cooperation between the two countries in three main areas. One of the areas was to further develop REDD+ demonstration activities by providing assistance that enabled activities 'to trial approaches to reduce emissions from deforestation and forest degradation [hence] KFCP was launched as a bilateral project between the Australian and Indonesian governments' (Australia Indonesia Partnership (AIP) 2009, p. 1). The other two areas the Australia Indonesia Partnership (AIP) cooperated on included 'policy development and capacity building to support participation in international negotiations and future carbon markets [and three], technical support for Indonesia to develop its national forest carbon accounting and monitoring system' (AIP 2009, p. 1). The project officially came to an end in June 2014 (D Central Kalimantan 2014), one year after a report noted on the Australian government's AusAID website that the programme was closing. 'The Commonwealth government has confirmed a $47 million project to restore 25,000 hectares of peatland on the Indonesian island of Kalimantan will end before most of its major milestones are met' (Arup 2013).

The 'early' closure of the REDD+ pilot project was unsurprising for many of those interviewed for the study, 'because it was like [a] public secret that it [KFCP] was failed' (SND Jakarta 2014). For other interviewees in Central Kalimantan, this perceived 'failure' was not a total failure as suggested, but nevertheless a self-fulfilling prophecy with deep irony. That is, the only objective that continued for a year after the closure was announced was the livelihoods stream. In an area wracked with poverty KFCP was for many of the participating villagers, adat women in particular, one of the few opportunities for employment through which they could at least alleviate their poverty (RD Central Kalimantan 2014; D Central Kalimantan 2014; M Central Kalimantan 2014).

The perceived failure of the KFCP focused on, but was not limited to, disagreements between adat communities excluded from and included in the

108 *Findings of the Indonesian study*

project (M Central Kalimantan 2014). Additionally, there was pervasive undermining of the project by a range of external and internal NGOs condemning, in particular, the FPIC process necessary for the project's transparency and efficacy (RD Central Kalimantan 2014). However, the respondents interviewed in Central Kalimantan, working with the project and its stakeholders from the time on-the-ground demonstration activities were implemented in 2010 to its closure in 2014, spoke to a more nuanced and pragmatic position about the merits of KFCP, and of its failures, perceived and real. The study's analysis identified this group as 'the pragmatists'. That is, they support the project in principle for its potential benefits of improved environmental and livelihood outcomes. However, this support was contingent on 'No Rights, No REDD' being upheld (D Central Kalimantan 2014).

The other participant groupings were located in Central Sulawesi and in Jakarta. Broadly, the participants in these groups were identified as 'the sceptics' and 'the globalists' respectively. These terms may appear overly reductionist; however, a deconstruction of the internal voices they represent proved analytically useful in adopting these descriptors to capture their views about REDD+. Finally, irrespective of the broad-based provincial descriptors I have applied to these groupings, one attitude was unanimously shared by all, and this was a deep reservation about REDD+ and how it would impact further on their already vulnerable communities. This deep reservation about REDD+ is primarily rooted in a history of national and international project interventions and unfulfilled promises by previous actors. That is, scepticism about the project was not only reserved for international actors but for the Indonesian government and its ministries at various levels including national (nasional), provincial (provinsi), district (kabupaten) and sub-district (kecamatan) levels. Both the multi-layered Indonesian governments and project facilitators in their conduct of 'development projects' obviously have, in the end, a final impact at the village (kampung) level, where projects such as REDD+ are implemented. The participants who fall into the three groups are, in some capacity or other, advocates, monitors, affiliates or representatives of adat communities. As such, the reservations they share are, in equal measure, representative of those expressed and felt by villagers at the local level.

The position of participants in Central Kalimantan is in no small measure connected to historical injustices wrought on the participating villages in the KFCP site area for demonstration activities. The designated area for the REDD+ pilot project was located within the Ex-Mega Rice Project (EMRP) area. Under the former Suharto regime, 1 million hectares of peat swamp forest were cleared from 1996–1998 in Central Kalimantan. The rationale for the peat forest deforestation provided at the time was to enable Indonesia to become self-sufficient in rice production through the conversion of what was regarded as largely unproductive peat swamp forest. An extensive network of irrigation canals was dug and the peat swamp drained. In spite of warnings from environmental groups and scientists that the project was doomed to fail, and an adverse environmental impact assessment that indicated such an outcome, the

Findings of the Indonesian study 109

project proceeded (RD Central Kalimantan 2014). Two years later environmental catastrophe followed.

> RD: Yeah the changing and then 1997 it was El Nino and then the area was dry and the peat land like if you want to have really good peat swamp forest you have to have better hydrological system and if you destruct the hydrological system. AG: Which the canals did. RD: Yes, it drained the land and it's like easily to catch the fire because of the fuel. AG: And then what happened after EMRP, so you were saying earlier when [President of Indonesia] Suharto fell in 1998. RD: Yes, in 1998. AG: There was a suspension of this logging related to Ex-Mega Rice Project. RD: Yeah when the EMR established like there were transmigrants moved to the area provided with housing and then per-diem, we call it per-diem, living cost, until they could be productive but the project failed and stopped. The fires occurred in 1997/8. AG: Is that the first time the fires were ... RD: no previously there were always fires like during crisis. AG: On this scale. RD: Like on this scale, huge scale that was the first like really big scale of fire, peat fire like forest occurrence in the area. There was always fire, but it wasn't like really big. AG: It was exacerbated by the El Nino and the dry conditions. RD: Yeah, yeah. And then caused like problems to the social in terms of the livelihood and their connection to the area and then also, livelihoods in terms of forest products, gardens and like you know the timbers, medicine ... all that had been provided by the forest is gone for their life.
>
> (RD Central Kalimantan 2014)

The unspoken agenda of Suharto's EMRP noted by the above participant, was never about self-sufficiency in rice production, it was designed to gain access to the precious timber inside the peat swamp forest. The extensive drainage canals dug allowed passage to access the timber and, hence, the ease with which the timber could then be transported out on to the Kapuas River, the main transportation artery in the KFCP area.

One decade following the failed EMRP, the KFCP project was launched with great fanfare by former Foreign Minister Alexander Downer in 2007 (Arup 2013). There was an underlying feeling of déjà vu for many, as *the* solution, to the ill-gotten social, economic and environmental problems that the EMRP had left behind. As one of the globalists explained, REDD+ must necessarily be seen within the purview of the effects of economic globalisation.

> So looking at the REDD, that for us at [name of NGO deleted] we see that as another development project, it's not the spirit, it's the same like ... *we are forced* to be convinced this is the solution for whatever problems that we are having right now. What environmental problem which is a [inaudible] problem.
>
> (D Jakarta 2014)

110   *Findings of the Indonesian study*

Overwhelmingly, for participants in the globalist group in Jakarta, KFCP appeared doomed to fail before it even commenced on-the-ground demonstration activities in 2010. The reasons for its anticipated failure were due to a range of factors, not least of which was the scale of the project's objectives which was considered to be too ambitious (RD Central Kalimantan 2014), and the socialisation activities, such as FPIC that preceded on-the-ground activities as lacking in both efficacy and transparency. One of the pragmatists in Central Kalimantan bluntly rejected the claims about FPIC, insisting that consent must necessarily be viewed as a work in progress, not something gained in one-off or even in several meetings to socialise participants about the project and tick appropriate boxes indicating consent had been sought and gained. FPIC is, was and always has been ongoing (RD Central Kalimantan 2014).

However, much of the strident criticism levelled at KFCP, by a range of international and local NGOs, targeted FPIC from its very early stages (Friends of the Earth et al. 2013). The principle of FPIC is rooted in the United Nations Declaration of the Rights of Indigenous Peoples and was contained in the KFCP Project Design (AIP 2009). It was not the principle the pragmatists objected to, but a single-minded obsession with FPIC by these groups that fundamentally blunted otherwise worthwhile aspects of the KFCP agenda. This group viewed FPIC as an ongoing process where trust was necessarily built over the progressive rollout of the project. For them, tangible issues were obscured by constant FPIC criticism, such as work on project activities which materially provided a much-needed source of household income to the adat communities involved in the project, as well as the potential to rehabilitate their severely degraded environment as a result of the environmentally disastrous and failed EMRP. Sometimes anger, but mostly frustration, was palpable with some participants in Central Kalimantan at what they considered was a constant stream of negativity about KFCP from inception, most of which was instigated by international environmental NGOs and their local (Indonesian) paid representatives. The projected failure of the pilot ultimately became a self-fulfilling prophecy, when the project was abandoned by the Australian government and the KFCP officially closed at a meeting in Jakarta in June 2014 (RD Central Kalimantan 2014).

Perhaps one of the greatest contributions to the Indonesian KFCP's anticipated failure was a wary – indeed weary – scepticism born of a history of big claims and broken promises by both national and international actors' project interventions. That is, the slow erosion of trust started with the promise of big agricultural cultivation with the EMRP, followed by international project interventions by NGOs boasting claims of rehabilitating the peatland and restoring livelihoods. One such project, initiated by the Netherlands government in 2006 and closed in 2008, was the Central Kalimantan Peatlands Project (CKPP). CKPP was a collaborative project made up by a number of local and international groups including Borneo Orang-utan Survival (BOS Mawas), CARE International, the University of Palangka Raya (UPR), Central Kalimantan (UNP), WWF and Wetlands.

Findings of the Indonesian study 111

CKPP was not dissimilar in its objectives to the KFCP, with the notable exception that the latter was designed as a climate change mitigation project in readiness for the international carbon market. That is, the objectives of CKPP included conservation, rehabilitation of the peatlands hydrology, canal blocking and empowerment of communities. BOS Mawas focused on rehabilitation. Wetlands were involved in canal blocking. CARE International's activities related to empowerment and livelihoods of adat communities, and UPR's involvement focused on research. WWF's work, while not in the Kapuas district, focused on the Sebangua area and was also involved in canal blocking and empowerment. Criticism of the project focused on poor coordination and communication 'in terms of ... interaction among the consortium itself. But it also didn't like have an impact I would say. We just like kept too much promise and the project finished' (RD Central Kalimantan 2014). With the closure of CKPP in 2008, soon after KFCP was launched in June 2009. A revolving door of international interventions, often with the same project personnel employed, drew fierce criticism by adat NGOs claiming '[t]he community sees the same CKPP personnel being engaged for the KFCP with a different project tag and hence are not convinced that KFCP will be any improvement from the previous CKPP' (Yayasan Petak Danum Kalimantan Tengah 2011).

When KFCP commenced on-the-ground activities in 2010, it was not long before a range of criticisms such as those expressed in the foregoing began to surface. In November 2010, the Executive Director of Yayasan Petak Danum Kalimantan Tengah (YPD) (Water Land Foundation Central Kalimantan), a collaboration of local NGOs which 'was kind of the hub for externalities to access the information about KFCP or to access the area of KFCP' (RD Central Kalimantan 2014), visited Canberra in Australia, raising a number of issues about the project. Following an Australian Delegation to Central Kalimantan in February 2011 to Kapuas district where KFCP was being piloted, YPD, on behalf of community leaders of the network, wrote a letter to the Delegation outlining their concerns. Included in a list of nine issues, number eight stated that YPD had '**no confidence in the international NGOs contracted to implement the pilot project**' (bold in original letter) (YPD 2011). Further, it is worth restating in full here, for the subtext of frustration underlying the lack of self-determination for adat communities:

> The engagement of existing international NGOs with working experience in the area by the Australian Government assumed a positive working relation with local communities. In the case of Borneo Orang-utan Survival (BOS), the relationship has been strained and estranged due to BOS' complete disrespect for the Dayak rights to the remaining forests which they have claimed as conservation area, without consultation with local communities for orang-utan rehabilitation. Both of the organisations [BOS and CKPP] have not had a track record of success in their previous restoration and conservation work, beyond carving out indigenous land and forests in the name of conservation which are out of bounds for the

112 *Findings of the Indonesian study*

Dayak. The community is not confident that the NGOs have the skills or the relevant experience to carry out environmental restoration or any other project activities in the area, beyond being paid personnel of the project.

(YPD 2011)

BOS's contract was discontinued when KFCP only maintained the livelihoods programme (RD Central Kalimantan 2014). However, the issue besetting all others, irrespective of national or international project intervention in the area, from EMRP to BOS Mawas, the CKPP's collaborators including BOS and then KFCP, is the lack of secure legal tenure for the Dayak communities, indeed for customary communities throughout Indonesia that exists to this day. This was a recurring theme for participants across all sites of the study – the lack of recognition and respect for customary (adat or indigenous) rights, which was also listed by the YPD as one of their major concerns to the Australian Delegation. YPD further argued that Australia's collaboration with the Indonesian government through the KFCP project 'inevitably condones this lack of consideration and hence the continued undermining of our customs and rights' (Yayasan Petak Danum Kalimantan Tengah 2011).

In an informal conversation with one Dayak participant from Central Kalimantan (where all participants were Dayak), some of the local NGOs listed as having endorsed the letter had no knowledge of its existence, while those who knew of it claimed it was conceived by Jakarta-based NGOs. Notable for their absence as endorsees on the letter was AMAN. That is, AMAN is the predominant independent social organisation of adat peoples throughout Indonesia, including Central Kalimantan.

With respect to recognition of adat rights and tenure, irrespective of international or national project interventions, securing these is, and has always been, a matter for the Indonesian government to put in appropriate regulations to effect this (RD Central Kalimantan 2014; D Central Kalimantan 2014; M'y Central Kalimantan 2014; M Central Kalimantan; the globalists group Jakarta 2014; the sceptics group Central Sulawesi 2014). As one participant claimed,

whether there is climate change or there is no climate change issue, it is not the thing that [is] needed by the adat communities. What is needed by the adat communities is the recognition, the protection by law about their activities on the ground. So you know they won't be impacted by any intervention.

(D Central Kalimantan 2014)

There has been some progress made on the issue of legal recognition. That is, in a landmark verdict by Indonesia's Constitutional Court in May 2013 the court ruled that indigenous peoples' customary forests should not be classed as state forest areas, thus confirming the existence of indigenous-owned customary forest. The relevant law that the Constitutional Court ruled as

Findings of the Indonesian study 113

unconstitutional, was enacted in 1999, Law Number 41 on Forestry (known as the Forestry Act). In presenting the facts of the case, the petitioners, one of whom was AMAN, stated that the Forestry Act was

> used as a tool by the state to take over the rights of indigenous people over their customary forest areas to become state forests, which were given/or handed over to capital owners, through various licensing schemes to be exploited without the consideration to the rights and local wisdom of indigenous peoples in the region, this had led to conflict between indigenous peoples with entrepreneurs exploiting their customary forest. Such practices occur in most parts of the Republic of Indonesia, which ultimately led to the rejection of the Forestry Law.
>
> (Decision Number 35/PUU-X/2012 (35/2012))

Decision 35/2012 was a major victory for adat communities; however, participant after participant interviewed in the Indonesian study reinforced that the ruling has had no effect on the ground. What is now needed to enforce the decision, they maintained, are regulations implemented at the national, provincial, district and sub-district levels of government.

As the first REDD+ demonstration activities pilot project in Indonesia, it is perhaps easier to comprehend KFCP's ostensible failure when legal recognition of adat communities' rights to their lands is viewed within an historical legacy of denial and delay by the state. However, each legal decision and regulation, as yet to be implemented, adds another layer to this palimpsest of adat recognition. Within this context, it appears particularly naïve to categorically state, as the YPD (2011) letter of dubious endorsement claimed, that Australia's KFCP is complicit by project association for an omission by the Indonesian state that bears decadal historical proportions, rooted in constitutional and regulatory impediments that can only be legislated by the state.

At the very least, what the detractors may wish to concede about the project, is partial acknowledgement that KFCP's presence, though connected to a fraught, problematic and protracted FPIC, has elevated the very issue that *all* participants in the study acknowledge impedes, and will continue to impede, any international or national interventions, and that is the lack of formal adat legal recognition. Indeed, one respondent confirmed in the context of legal recognition that 'REDD it is already talking about the issue of adat communities, indigenous communities he said, adat communities that live inside the forest' (D Central Kalimantan 2014). This acknowledgement made by a locally based NGO Dayak participant, confirms what the project design for KFCP perceived its role to be as an 'instrument of change'. That is,

> Clear land tenure laws cannot be made a precondition of project development, because no projects would then ever be developed or they would all be developed in the same handful of places. Rather, the projects themselves can be made the instrument of change, where community

114   *Findings of the Indonesian study*

> management rights are first given to local people in a step-wise process to full land tenure.
>
> (AIP 2009)

However, in a joint press release issued by YPD, WAHLI (Friends of the Earth, Indonesia) and Friends of the Earth Australia, there appears to be no acknowledgement that, indeed, KFCP has played a contributing role in this important end goal. 'Customary landholders deserve a better deal than what they are currently getting ... [s]upporting recognition of their rights is a way of countering these destructive industries [oil palm plantations, logging concessions, mining], and investing long term in the conservation estate' (Friends of the Earth International et al. 2013).

The lazy conflation of KFCP with the activities of destructive industries simply confuses the core component of demonstration activities of the project – that is, restoration of deforested and degraded peat swamp forest (AIP 2009, p. 5). There is no disagreement from the participants that the real drivers of deforestation and forest degradation in Indonesia are indeed attributable to these industries (N Jakarta 2014; D Jakarta 2014; SND Jakarta 2014; A Jakarta 2014). However, the constant stream of criticism and negativity, from both 'outsiders' and 'insiders' undermining the project, has had its part to play in the project's 'failure', which ironically has allowed the very industries that cause much of forest destruction in Indonesia, for example, oil palm plantations, to take up its place, 'as they are now doing' (RD Central Kalimantan 2014).

As a final point, it is particularly condescending to imply that 'bules' (foreigners) through their project presence can effect change of this scale and significance when the agency of Indonesian NGOs interviewed in the course of this study are more than capable, have been and continue to be so in the carriage of this long fought and ongoing battle for adat recognition. Indeed, other victories of constitutional significance preceding the landmark Decision 35/2012 was Decision 45/2011 where the Constitutional Court removed the state's authority to unilaterally establish forest status (National Forestry Council People's Chamber & HuMa 2013). Decision 45/2011 was the result of five district heads in Central Kalimantan challenging the Forestry Act that designated their administrative districts as part of the forestry zone. This affected some 130 million hectares of Indonesia's Forest Zone (Lang 2013). However, in spite of significant victories having been won, as one participant working on recognition noted, Decision 35/2012 only gives 'jurisdictional recognition' and that is not enough. 'Regulation to implement the decisions is what is now needed' (D Central Kalimantan 2014).

When asked what the outcomes of KFCP were at the meeting in Jakarta, one respondent who was present at the June 2014 closure meeting of the project, who had not been employed by the project's organisational structure but who had conducted extensive research with the participating Dayak communities, categorically stated 'confusing':

Findings of the Indonesian study 115

You've got to sort of like have a good understanding ... about this first before it's taking place [KFCP]. Because like you said before, what is the outcomes of this is. Confusion. Confusion is the outcomes until today. Because so many, it's very easy for like you know the externalities with the different agenda to criticise this project.

(RD Central Kalimantan 2014)

Probing who the 'externalities' were, the respondent claimed organisations outside the area. 'Meaning national, international that had no connection with the area and [a] lack of understanding about local context' (RD Central Kalimantan 2014). Adding that perhaps the ambitious scale of the project made it susceptible to scrutiny. The project design for KFCP stated that,

KFCP's objective is to demonstrate a credible, equitable, and effective approach to reducing emissions from deforestation and forest degradation, including from the degradation of peat lands, which can inform a post-2012 global climate change agreement. As part of this, the KFCP aims to trial a range of approaches to show how investment in REDD can achieve emission reductions while providing forest-dependent communities with livelihoods and promoting sustainable resource management. It will also contribute to developing governance, enforcement, and regulatory frameworks to support REDD. Lessons learned from the KFCP will help demonstrate how REDD can be part of a post-2012 global climate change agreement and how the approaches and methodologies tested in Central Kalimantan can be scaled up or replicated in other parts of Indonesia.

(AIP 2009, p. 12)

Central Kalimantan participants shared a common theme in their assessment of KFCP, irrespective of its ambitious scale as a demonstration pilot of how a global-scale incentive-based mitigation programme under the name of REDD+ may work or not. And that was, it provided a much-needed source of household income in an ecologically damaged environment. In the opinion of participants, KFCP's emphasis on the environmental outcomes, specifically related to carbon mitigation, was too narrow and misguided, claiming equal weight should have been taken into consideration about the socioeconomic needs of the participating villages. In the opinion of one participant, KFCP was never about carbon mitigation for the participating villages; it was about livelihoods (RD Central Kalimantan 2014).

Until now, the focus has been on Palangkaraya, Central Kalimantan and Jakarta participants' responses in these general findings to the exclusion of Central Sulawesi's. In part, this is due to the UN-REDD+ pilot under study, the KFCP, having come full circle. In other words, when interviewing commenced in June 2014 in Palangkaraya, this coincided with the closure of the KFCP pilot project. In many respects and in hindsight, the timing presented a unique opportunity to interview participants who were able to assess the

116 *Findings of the Indonesian study*

project over its lifespan and, in so doing, their responses were reflective in nature. That is, there was a level of participants' self-reflection on KFCP's promises and those that were not fulfilled, its faults, its potential, and its perceived and real failures.

The focus now turns to the limited interview data I gained from the Palu participants in Central Sulawesi. 'Limited' not in substantive data insights gained, but due to issues regarding the lack of English translators I encountered upon arrival. Notwithstanding this issue, central themes and patterns that emerged in Central Kalimantan may be drawn.

Outcomes and analysis – Central Sulawesi

Central Sulawesi, in contrast to Central Kalimantan, had yet to commence on-the-ground demonstration activities at the time of my arrival and departure, mid- to late June 2014. That is, REDD+ readiness activities, the pre-implementation phase, namely socialisation and FPIC, had just been completed at the time participants were interviewed. In many ways, this presented the opportunity to examine, in a cross-provincial way, whether or not the outcomes of the KFCP in Central Kalimantan were evident in REDD+ Central Sulawesi at the end of their socialisation phase. That is, what the similarities or differences between the two provinces were. However, when asked whether or not they were familiar with REDD+ activities, the answer was negative. Indeed, as one informal conversation suggested, Central Kalimantan was up for any outside intervention as this had been part of their history, therefore REDD+ was no different. The only commonality existing between respondents in Central Kalimantan and Central Sulawesi were organisational affiliations with Jakarta-based research partner, HuMa. Thus, to speak of research outcomes in Palu is a synthesis of data gathered and analysed from both sites, that is, Jakarta and Palu.

In October 2010, the province of Central Sulawesi was selected by the UN-REDD+ Programme as the focus site for Demonstration Activities in preparation for REDD+ implementation. Five districts (kabupaten) were nominated as sites for REDD+ demonstration activities: Sigi, Donggala (Dampelas region), Parigi Moutong (Tinombo region), Tojo Una and Toli-toli (FPP 2011). To coordinate the demonstration activities, a REDD+ working group was established, Pokja, comprising representatives of government and NGOs as well as the Indonesian Association for Forest Corporation (PA1 Central Sulawesi 2014).

Similar responses in Central Kalimantan emerged in Central Sulawesi in perceiving REDD+ as just another development project that follows a long history of intervention by the Indonesian state, in particular with regard to concessions granted to mining, logging and palm oil companies. With respect to REDD+, an overriding concern expressed was the 'social changes to the community' once REDD+ is implemented (AP1 Central Sulawesi 2014). When probed about what social changes were anticipated, the respondent's hope was

Findings of the Indonesian study 117

that REDD+ would decrease the poverty in the communities in which REDD+ is implemented. Improving livelihoods and alleviating poverty, however, were linked to maintaining access to ancestral lands, with a great deal of scepticism, hence the provincial term I applied to those interviewees in Palu, Central Sulawesi as 'the sceptics'.

> PA: It is the same trick between how they make a programme for poverty in the city for people in the forest. Government think that the poverty in the forest is the same like the people in the city. They think, for example, they think the people in the forest they don't have rice so they send the rice to people in the forest, for me it is like a very ridiculous programme because the people always have rice ... AG: What's the poverty in the forest? PA: The people in the forest [from the national government's perspective] we are category, we are poor because they see we have no TV, have bad house, but there is not the poverty for them, for example, can't access ... So people in the forest their poverty is because of their access [to the forest]. S/He say they have long history, how they life connected with the forest and that history the fact is as long as their life they don't have deforestation with the forest, they're like parents their life with the forest ... so s/he say agenda of conservation coming to people in the village in the forest, only two words, two words, 'forbidden' and 'prohibit'.
>
> (PA2 Central Sulawesi 2014)

The agenda for REDD+ may not explicitly mention 'forbidden' and 'prohibit', but to the same respondent, 'REDD is like durian candy [the number one fruit in Indonesia] that presents as all about profit but what comes after it is just the same'. The 'same' is, 'don't let it drop in the same hole' (PA Central Kalimantan 2014).

The central problem in Central Sulawesi is the universal claim of customary communities in Indonesia: the continued lack of legal recognition of customary forests. Like their counterparts in Central Kalimantan, *regulation to implement* the Constitutional Court ruling remains outstanding at the sub-national level, that is, the provincial government level.

Other similarities in the REDD+ socialisation phase that existed between the two provinces under study returned to the issue of FPIC. Central Sulawesi respondents' claims of elite-level language and the lack of information to make decisions about REDD+ was a common theme (PA Central Sulawesi 2014; M Central Sulawesi 2014; SB Central Sulawesi 2014). One respondent claimed the language about REDD+ was not adapted to a level understandable at the village level. 'Although [in] the village, it's still with the elite language. It's about the people of the team that go to the village, because in the experience of [name deleted] village the people kick out the team' (M Central Sulawesi 2014). And then, there is the problem of what or who constitutes indigeneity. This latter distinction only emerged in participants' responses in Central Sulawesi.

Findings of the Indonesian study

Conclusion

The aim in presenting the Indonesian study's general findings mediated through the direct voices of the study's participants, is to provide a platform, unvarnished by author privilege of authentic on-the-ground views on the projects under study. The following chapter is more theoretically interventionist in its critique of sub-surface meanings offered in the responses.

The most salient general finding to emerge from participants in both Central Kalimantan and Central Sulawesi is the poverty, and necessity, of adat recognition via secure tenure that requires implementation of the Indonesian Constitutional Court's landmark ruling.

References

Arup, T 2013, 'Australia-Indonesia carbon project abandoned', *The Sydney Morning Herald*, 22 July, accessed 17 June 2017, www.smh.com.au/federal-politics/politica l-news/australianindonesian-carbon-project-abandoned-20130702-2p98w.html.

Australia Indonesia Partnership (AIP) 2009, *Kalimantan forests and climate partnership (KFCP) design document*, Indonesia-Australia Forest Carbon Partnership, accessed 15 January 2015, http://formin.finland.fi/public/download.aspx?ID=48885&GUID= %7B9B0BA3BA-25BF-4FEA-985B-B6DADCA60EAC%7D.

Barnett, J & O'Neill, S 2010, 'Maladaptation', *Global Environmental Change*, vol. 20, pp. 211–213.

Down to Earth 2013, 'A turning point for Indonesia's indigenous peoples', *Down to Earth*, 13 June, accessed 16 October 2020, www.downtoearth-indonesia.org/id/node/1025.

Forest Peoples Programme (FPP) 2011, 'Central Sulawesi: UN-REDD Indonesia's Pilot Province', *Forest Peoples Programme* Rights, forests and climate briefing series, October 2011, accessed March 2013, www.forestpeoples.org/sites/default/files/publica tion/2011/10/central-sulawesi-briefing-4.pdf.

Forest Peoples Programme (FPP) 2013, 'Constitutional Court ruling restores indigenous peoples' rights to their customary forests in Indonesia', *Forest Peoples Programme*, 16 May, accessed 16 October 2020, www.forestpeoples.org/en/topics/rights-land-natural-resources/news/2013/05/constitutional-court-ruling-restores-indige nous-pe.

Fraser, N 2008, 'Reframing justice in a globalizing world', in N Fraser, *Scales of justice: reimagining political space in a globalizing world*, Polity Press, Cambridge, UK.

Friends of the Earth International, Friends of the Earth Australia, Yayasan Petak Danum Banjarmasi, WAHLI/Friends of the Earth Indonesia 2013, 'Friends of the Earth calls for transparency on climate aid and recognition of customary land rights in Indonesia', *REDD-Monitor*, accessed 20 May 2017, www.redd-monitor.org/2013/08/21/friends-of-the-earth-calls-for-an-open-review-of-the-kalimantan-forest-carbon-p artnership/#more-14283.

Juhola, S, Glaas, E, Linner, B & Neset, T 2016, 'Redefining maladaptation', *Environmental Science & Policy*, vol. 55, January, pp. 135–140.

Lang, C 2013, 'Indigenous peoples' rights and the status of forest land in Indonesia', *REDD-Monitor*, weblog post, 9 October, accessed 10 May 2017, www.reddmonitor. org/2013/10/09/indigenous-peoples-rights-and-the-status-of-forest-land-in-indonesia/.

Magnan, AK, Schipper, ELF, Burkett, M, Bharwani, S, Burton, I, Erikson, S, Gemenne, F, Schaar, J & Zievogel, G 2016, 'Addressing the risk of maladaptation to climate change', *Wiley Interdisciplinary Reviews: Climate Change*, vol. 7, September/October, pp. 646–665.

Maxton-Lee, B 2018, 'Narratives of sustainability: a lesson from Indonesia: global institutions are seeking to shape an understanding of sustainability that undermines its challenge to their world view', *Sounding: A journal of politics and culture*, no. 70, Winter, pp. 45–57.

Maxton-Lee, B 2020, 'Forests, carbon markets, and capitalism: how deforestation in Indonesia became a geo-political hornet's nest', Guest Post, *REDD-Monitor*, accessed 22 August 2020, https://redd-monitor.org/2020/08/21/guest-post-forests-carbon-markets-and-capitalism-how-deforestation-in-indonesia-became-a-geo-political-hornets-nest/.

National Forestry Council People's Chamber & HuMa 2013, 'Indonesia: the government must step up action to meet the constitutional rights of indigenous people and local communities as part of the process of determining the status of forest land', *Indigenous Peoples Human Rights Defenders Network (IPHRD)*, accessed 10 June 2017, http://iphrdefenders.net/indonesia-the-government-must-step-up-action-to-meet-the-constitutional-rights-of-indigenous-people-and-local-communities-as-part-of-the-process-of-determining-the-status-of-forest-land/.

Oliver-Smith, A & Shen, X (eds) 2009, *Linking environmental change, migration and social vulnerability*, UNU Institute for Environment and Human Security (UNU-EHS), e-book, accessed 21 May 2017, www.munichre-foundation.org/dms/MRS/Documents/Source2009_OliverSmith_ShenEnvironmentalChange_Migration.pdf.

Röhr, U 2006, 'Gender relations in international climate change negotiations', *Berlin: LIFE eV/genanet*.

Schipper, ELF 2006, 'Conceptual history of adaptation in the UNFCCC process', *Review of European Community & International Environmental Law*, vol. 15, no. 1, pp. 82–92.

Schipper, ELF 2009, 'Conceptual history of adaptation in the UNFCCC process', in ELF Schipper & I Burton (eds), *The Earthscan reader on climate change*, Earthscan, London, pp. 359–376.

The Forests Dialogue 2012, *Background paper for the scoping dialogue on the exclusion and inclusion of women in the forestry sector*, Women Organising for Climate Change in Agriculture and NRM, accessed 10 June 2017, http://theforestsdialogue.org/publication/background-paper-scoping-dialogue-exclusion-and-inclusion-women-forestry-sector.

Work, C, Rong, V, Song, D & Scheidel, A 2019, 'Maladaptation and development as usual? Investigating climate change mitigation and adaptation projects in Cambodia', *Climate Policy*, vol. 19, no. 51, pp. 547–562.

Yayasan Petak Danum Kalimantan Tengah (YPD) 2011, 'To the Australian delegation to Central Kalimantan February 2011, re: community concerns with the KFCP', *REDD-Monitor*, accessed 20 May 2017, www.redd-monitor.org/wp-content/uploads/2011/02/YPD-Letter-to-Australian-Delegation.pdf.

7 Where are the women?

By chapter's end, 'Where are the women?' is a question anticipating their ongoing exclusion; a questioning point of priority and reflection for decision makers and stakeholders involved in the United Nations Programme on Reducing Emissions from Deforestation and Forest Degradation (UN-REDD+), at all levels and spatial scales. The Center for International Forestry Research's (CIFOR) review of the first 10 years of REDD+ found it gender blind, with a systemic disinterest about gender issues in general in implementing countries of the global South (Angelsen et al. 2018, cited in Global Forest Coalition 2020). The Global Forest Coalition (2020) published a 15-year review of REDD+ producing similar results. Gender mainstreaming, as the main mechanism through which to include women in REDD+'s participatory processes, amounts to procedural inclusion, for instrumental ends. Namely, where women and customary communities are 'silent offsets' under the global mitigation programme of REDD+. By this account, social reproduction is leveraged to further the strategic interests of carbon capitalism, thus giving rise to the notion of a gendered silent offset economy.

Introduction

> with so many demands from civil society to put women everywhere. Putting women everywhere does not mean that you have mainstreamed them into the heart of the policies.
>
> (AP2 Jakarta 2014)

This chapter returns to the invisibility of women in climate change discourse discussed previously and juxtaposes these insights and ideas with findings revealed from on-the-ground REDD+ projects and readiness activities in the Indonesian study. By presenting this dedicated chapter on women-specific findings, the aim is to avoid replicating any gender omission/gap/missing link/silence or blindness in the discourse on climate change solutions by providing a clear, visible and voluble account of adat women's understandings about three climate change dimensions the study investigated. Namely, the 'gendered impacts of UN-REDD+' on adat women specifically, 'carbon markets' and 'climate/gender justice'.

The first section merges the outcomes of the qualitative study and their discussion with insights from the book's earlier conceptual chapters to facilitate a more rigorous analysis of the study's empirical outcomes. Following this analysis, the concluding section will be in a position to answer the question of this chapter's title: Where are the women?

Preface to findings and discussion

One of the major themes to emerge across the three research sites of the study was the issue of securing masyarakat adat (customary community) tenure. The discussion of this general finding centred on the political and legislative complexities the historical claim entails. However, the focus on tenure issues had the effect of overwhelming other findings relating specifically to women. This is not to suggest adat women do not share the goal of securing tenure; they do. However, the predominance of this community priority had the tendency to overshadow their voice and their priorities in relation to projects such as REDD+. This accords with Arora-Jonsson's (2014) research reviewing 40 years of women and environmental policy where 'prescriptions on the importance of market-based programs often disregard women's own preferences and understanding of their livelihoods' (Arora-Jonsson 2014, p. 302). 'Who is telling the story?' is one of the enduring myths in forest governance discourse, where the 'who' has the capacity to reinforce existing hierarchies, while marginalising dissenting and critical voices, which is crucial to its perpetuation (Delabre et al. 2020, p. 1). The aim here is to give salience to what adat women's subjective needs and priorities are. As importantly, what structural impediments hinder their realisation.

Finally, a note on the subject of how participants responded to women-specific questions in the semi-structured interviews is foreshadowed here. What was notably lacking were specific references to women in responses, in spite of questions being framed around adat women's understandings of the thematic areas under study. Eliciting this information required, in many cases but not all, further probing and/or rephrasing of the questions. For example, responses from questions seeking gender disaggregated data under the theme of gendered impacts – 'What are the impacts of REDD+ on perempuan adat (customary women)?' – were frequently answered with the use of a collective pronoun, such as 'household', 'family' or 'community'. As one respondent stated, this is due to the fact that 'they [Kalimantan Forests and Climate Partnership – KFCP] see the target, they make it a target ... not men or woman not a child or older, just a family [household]' (M'y Central Kalimantan 2014). Women Organizing for Change in Agriculture and Natural Resource Management (WOCAN 2012) say a plethora of recommendations exist to improve the inclusion of women at a local level, in the context of forestry, for example, 'governance structures of community management schemes should define membership as open to all adults, both women and men, as opposed to using the household single unit' (FAO 2007, Sarin 1995, cited in The Forests Dialogue 2012).

122 *Where are the women?*

All of the participants who were interviewed had varying levels of expertise on climate change programmes through their organisations; however, the participants were not adat living in the sites where UN-REDD+ projects and readiness activities were conducted. In the case of the Jakarta group, the globalists, they comprised solely women participants who were employed by various non-governmental organisations (NGOs) representing masyarakat adat, and worked on issues that affected these communities, climate change included. Additionally, this group of participants had strong connections with international institutions and a well-established network of partner organisations in Central Kalimantan and Central Sulawesi, with one participant pointing out that Indonesian NGOs are 'all about networking' (AP2 Jakarta 2014). The participants interviewed from these provincial capitals, respectively Palangkaraya and Palu, were identified in Jakarta by research partner HuMa; all participants were adat and all had expert *local knowledge* about UN-REDD+ and its *in situ* impacts.

Findings and discussion of findings of the Indonesian study

Gendered impacts of REDD+ on women and gender mainstreaming

Under this theme, several sub-issues from the semi-structured questionnaire were investigated, often producing overlapping responses from respondents. Namely: assessing how adat women are 'gender mainstreamed' in the UN-REDD+ mechanism; examining the strengths and weaknesses of REDD+ for forest-dependent adat women and their livelihoods; and disaggregating the impacts of REDD+ on adat women and men involved in REDD+ socialisation and implementation activities. In questioning participants about how adat women are 'gender mainstreamed in REDD+', and how this term is understood by women who are engaged in on-the-ground REDD+ activities, responses were as follows:

> D: When we are talking about gender mainstreaming on the paper, on the documents it's totally 'gender-wash'. AG: In what way is it total 'gender-wash'? D: Talking about participation, they put the pillars there in many projects, development, of how do you say, discourse recently, but one of the pillars, always participation of the women, and they put like quantitative measures on that level you think and must be involved in this. How can you ensure that 30 per cent of women is there and how can you ensure women accommodated? It is just numbers.
>
> (D Jakarta 2014)

> No, no. They just put on paper [30 per cent of women participating in UN-REDD+]. In indigenous community also, they don't have to give some position decision that decide men and women, this makes no practical [inaudible]. I don't know if it's coming from our history, or sometimes

religion, [women] don't want make a decision. Woman must be behind the kitchen ... absolutely just a second layer.

(S Jakarta 2014)

Yeah and talking about participation and gender mainstreaming implementation of the document [REDD+], it's actually to just to speed up the process and present a report ... to the donors and the international world, like the UNFCCC [United Nations Framework Convention on Climate Change] like going to be submitted to the IPCC [Intergovernmental Panel on Climate Change] report and nothing there you know.

(N Jakarta 2014)

In Kapuas district is one of the site of REDD project and also KFCP project we are impart them [women] into the activity and we asked them – you know about the gender, you know about something with the gender mainstreaming there is no one can understand about that.

(M Central Kalimantan 2014)

Gender mainstreaming was established in Indonesia's National Development under Presidential Instruction No.9/2000 which provides 'a mandate for all government agencies to mainstream gender in their policies and programs' (WOCAN 2012, p. 2). With respect to gender mainstreaming in REDD+, Indonesia established a National REDD+ Task Force to develop its National REDD+ Strategy (STRANAS). The strategy document states that to ensure gender equality is achieved, a safeguard by implementing agencies of REDD+ is to be adhered to that includes conducting 'capacity building activities for women and other vulnerable groups to enable them to access and understand information, and meaningfully participate in decision-making processes' (WOCAN 2012, p. 2). Further, to ensure that women participate meaningfully and effectively in policy development and implementation of REDD+, the guidance note recommends that 30 per cent of women are to be included with an aim to achieve 50 per cent (WOCAN 2012).

The disconnect between the participant responses on gender mainstreaming REDD+ and Indonesia's national infrastructure to support its implementation, speaks of a wider normative disjuncture about women's inclusion in on-the-ground implementation activities with its theory. Originally adopted at the Fourth World Conference of Women in Beijing in 1995, gender mainstreaming was developed within a radical feminist framework by women's lobby groups and international NGOs at a transnational level. 'Yet translation of mainstreaming occurs at the local level where national contextual realities may clash with the original feminist conceptualisation' (Walby 2005, cited in Alston 2014, p. 289). The original conceptualisation of gender mainstreaming envisaged 'transformative' gender equality by eliminating the 'invidious invisibilities of sex class' (Zalewski 2010).

The stated purpose of mainstreaming gender in national policies and programmes, according to Indonesia's National Development, is to achieve

124 *Where are the women?*

'gender equality'. However, gender equality is narrowly interpreted in REDD+ by circumscribing it around, among others, 'meaningful participation'. Participation is further, and problematically, reduced to 'quantitative measures' and, as questioned by one of the participants, 'how can you ensure that 30 per cent of women is there?' (D Jakarta 2014). Another baldly stated women are gender mainstreamed in REDD+ as follows:

> If they need to bring women representative, then the head of the village will find someone from the community but it is only like just to fill out the decision to take the women representative.
>
> (A Jakarta 2014)

The respondents' criticisms are not new, but share an ongoing critique of gender mainstreaming by feminist scholars for its failure to deliver substantive changes on gender inequality since the 1995 Beijing Conference. In spite of its transformative promises, an early criticism of gender mainstreaming described it as an 'empty signifier' (Council of Europe 1998, cited in Alston 2014). More recently, its meaning has been reduced to a 'formalistic ritual' and 'technocratic exercise' (Arora-Jonsson 2014). That is, 'gender mainstreaming has become procedural (Meier and Celis 2011), in many cases a tick the box exercise where gender analysis is reduced to gender statistical analysis ... at the expense of transformative change' (Alston 2014, p. 290). Procedural gender mainstreaming in REDD+ then mimics the wider approach to environmental problems such as climate change, which are treated as technological and engineering problems, 'but are part of an ideational crisis' (Delabre et al. 2020, p. 1).

At the time gender mainstreaming was adopted as a major strategy to achieve gender equality at the 1995 Beijing Conference, critical voices were being raised, especially from women's groups in the global South, lamenting the fact that wherein the focus that had previously been on women, it 'shifted to women and men, for example, by promoting discussions on "men at risk"' (Baden & Goetz 1998). One could say that 'gender' became important, but women had disappeared (Arora-Jonsson 2014, p. 301).

Where are the women? As the earlier responses suggest, the adat women in the Indonesian study have not disappeared; they are there, but their visibility and voice have been diluted through a protracted battle for securing adat tenure, as the previous chapter's findings disclosed, and from 'mainstreaming gender' itself.

Given gender mainstreaming is the current strategy to achieve gender equality, where gender equality's implementation is to be mediated through 'meaningful participation' in REDD+, then laying bare the impediments hindering adat women's participation will, conversely, re-embody their presence. Some of the barriers to participation, 'meaningful' or otherwise, included culturally embedded and entrenched stereotypes/social roles, elite language and lack of education, expanded on shortly. These impediments have existed for decades and are congruent with 'reasons for women's exclusion' identified in the forestry sector in countries of the global South, including but not limited

Where are the women? 125

to: a form of role theory that prescribes the sort of work women and men can do based on the notion that 'roles are normative and express expectations of ideal behaviour'; which in turn feeds into culturally determined stereotypes that 'make it natural that they [men and women] do different jobs' (Gurung 2002, cited in The Forests Dialogue 2012). For example, as one respondent previously stated, '[w]oman must be behind the kitchen ... absolutely just a second layer' (S Jakarta 2014).

However, the attribution of women's subordinated status is not simply an artefact of culturally embedded stereotypes and social roles. One respondent suggested their exclusion is related to an additional and more deeply rooted psychological reason.

> Actually there is no self-esteem of the women to come to participation in some activity. Maybe they just push the men to get the activity because they have thinking about we are just women, we are just talk about the household, like that.
>
> (M Central Kalimantan 2014)

> sometimes the women are just on the second layer for making decisions and sometimes women just silent and they don't like to put their hand [up] but they want to talk with the decision ... Because you know a lot of women in Indonesia are silent and are sometime they're forced, sometimes scared to talk.
>
> (S Jakarta 2014)

These responses indicate that not only is it manifestly inadequate and cli_chéd to 'add women and stir' (Arora-Jonsson 2014), as a reading on procedural gender mainstreaming would have it; rather, they suggest that a necessary *precondition* for women to participate in REDD+ decision-making and implementation activities, in any 'meaningful' way, requires dedicated programmes to specifically address elevating adat women's sense of self-worth. This is not the interpretation of the interviewer and author of this book – a white, privileged, educated bule (foreigner) woman – but from the respondents themselves. That is,

> Before REDD there was not many talk about woman, and REDD established then, like you know, more and more discussion about like woman needs to be involved. Like you know they need to speak up they have to present themselves, their needs, their priorities ... because like you know the intervention itself much more focused on the community ... [what is needed is to] make them self-esteem, make them able to speak up, make them feel confidence about themself which ... will take time and efforts like that. You have to be focused on woman, like assist them to make them to be able to be on that capacity, to be on that level.
>
> (RD2 Central Kalimantan 2015)

126 *Where are the women?*

However, KFCP and other REDD+ projects are not designed to build women's self-esteem and confidence through psychologically enabling well-being programmes that would probably, in part, facilitate the 'meaningful participation' the policy document prescribes. In essence, REDD+ is simply part of a global abatement programme designed and structured to mitigate carbon emissions by reducing deforestation and forest degradation. The National REDD+ Strategy explicitly states that to ensure gender equality is achieved, implementing agencies of REDD+ are to adhere to a safeguard by conducting 'capacity building activities for women and other vulnerable groups to enable them to access and understand information, and meaningfully participate in decision-making processes' (UN-REDD Programme Indonesia 2012, p. 2). However, the Indonesian study found no evidence through interviews with participants that the REDD+ 'safeguard' of capacity-building activities extends to self-esteem programmes that would 'assist them [adat women] to be able to be on that capacity, to be on that level' (RD2 Central Kalimantan 2014).

Rather than change programme structures to accommodate the needs, ideas and subjective positions of the women themselves to redress the disadvantages impeding meaningful participation, however, 'women have often been expected to join organisations and accommodate themselves to existing norms and structures' (Arora-Jonsson 2014, p. 303). REDD+, in other words, like many of the other 'development' programmes previously discussed, is about integrating women into what are essentially male normative systems (Alston 2014).

> [Even though] most negotiators have been sensitised with so many demands from civil society to put women everywhere. Putting women everywhere does not mean that you have mainstreamed them into the heart of the policies. From my point of view for example, [at UNFCCC negotiations] you talk about MRV [measurement, reporting, verification], you talk about carbon measurement algorithm. It's really masculine don't you think?
>
> (AP2 Jakarta 2014)[1]

This response opens findings to one of the other major themes that the semi-structured questionnaire investigated, namely 'carbon markets' and adat women's understanding of these, addressed later. For now, the discussion of the gendered impacts of REDD+ continues, where there was invariably some level of cross-over in participants' responses to themes not viewed in isolation, but relationally.

Firstly, recalling several impediments hindering the meaningful participation of adat women intimated in the opening verbatim responses on gender mainstreaming: culturally embedded stereotypes/social roles, elite language and education. These inhibitors to participation in REDD+ decision making and activities were not confined to responses relating to the gendered impacts dimension of the semi-structured questionnaire but were present in responses regarding the two other thematic areas the questionnaire investigated. Namely, carbon markets and justice.

If 'gender is known by educated people' (M Central Kalimantan 2014), as one participant claimed, then not only is 'gender', let alone 'gender mainstreaming', misunderstood by women at the village (kampung) level, but more fundamentally,

> M: Actually in the field KFCP project or REDD project are very low understanding of the women about the gender *because mainstream gender does not really apply in the village.* AG: OK. M: And this for example, in [name withheld] is one of the site of REDD project and also KFCP project we are impart them into the activity and we asked them – you know about the gender, you know about something with the gender mainstreaming there is no one can understand about that.
>
> (M Central Kalimantan 2014, emphasis added)

The reason why gender/gender mainstreaming is not understood by the adat women is not a simple case of incomprehension, but more essentially, as the response indicates, because such concepts are culturally inapplicable. For example, in Central Kalimantan there are three major Dayak tribes, of which Dayak Ngaju and Dayak Ma'anyan are two. In the course of this research in Palangkaraya over two and half months and two visits, the author lived with the family of her translator, where the mother was Dayak Lamandau and her husband Dayak Ma'anyan. The mother fluently spoke not only her tribal Dayak Lamandau but also Dayak Ngaju, Dayak Ma'anyan, Bahasa Indonesian as well as sedikit Ingriss (a little English). She told me her father was the village chief in her kampung for many, many years, right up until the time he died. The community had built not one, but two long-houses (up to 25–30 metres in length) to honour him, not solely because of his position but rather because he was perceived to be a fair chief, a good protector, and an honourable and intelligent man. He was well respected and loved by all.

This personal insight is as touching as it is theoretically revealing. While the research has not used an anthropological lens to operationalise the in-field study, the substance of the story is, theoretically speaking, politically and culturally anthropological in nature. It provides a snapshot view and level of nuance rooted in the particularity of Dayak diversity. In this case, the Dayak Lamandau, who have their own traditions, notions of equality, ways of doing, living, being, communicating respect, and as importantly specificity of language. UN-REDD+ can never hope to capture the cultural heterogeneity and particular differences within and between the Dayak communities, nor was it designed to. Just as the Dayak in Central Kalimantan are not homogenous, so too

> women is not homogenous. But from this gender mainstreaming thing, they are just being clapped into one group. No reflection, no different access or no reflection of different access to information, even to analyse what is going on, they have just been invited.
>
> (N Jakarta 2014)

128 *Where are the women?*

Similarly, the transposition of a transnational strategy like gender mainstreaming in REDD+ in Indonesian Borneo to achieve gender equality begs another question – 'whose equality'? We know that critical voices from women of the global South at the Beijing Women's Conference in 1995 were decrying the displacement of the discussion on women then, with the depoliticised term 'gender' leading to discussions about 'men at risk'.

What do the women want from REDD+? The Ngaju Dayak women in the KFCP site want equality alright.

> Specifically, what the woman want to them in that area, we just ask the people, we just ask the woman and then what do you want if we talk about the climate change is we talk about the REDD ... we ask and then [they] ask something for health, ask to manage our family in this village and then, until now, KFCP did not give a response for that.
>
> (M'y Central Kalimantan 2014)

It is not as if women have not found their voice to express their needs and priorities when asked; they did and they have, but they have been ignored. While women have not been excluded in KFCP, their inclusion has been strategically enlisted in the service of meeting the priorities of the project, thus reaffirming Arora-Jonsson's (2014) observation that women are expected to accommodate themselves into existing norms and structures instead of the project structure being changed to suit their subjective positions.

What are the adat women's subjective positions? There was one area that the KFCP project document suggested would be 'advisable' to exclude women from, and this was engaging in activities in relation to peatland development, such as fire-fighting and land clearing, both of which have been traditional male tasks among Dayak Ngaju, the local tribe in Kapuas district. The advisability of women's exclusion from what is a core economic activity in the project was not solely based on the male-dominated nature of these tasks, but framed within a deeper understanding that '[t]he burden women are carrying is [already] significant' (AIP 2009, Attachments 1–3). The demands placed on adat women are not simply confined to their work burden which is often made up of labour-intensive activities that are usually less economically attractive, but also related to health issues, specifically maternal reproductive issues.

In the KFCP area, for example, fertility rates are high, maternal health is poor, underage marriages are prevalent, the latter of which results in relatively high mortality rates and low birth weights – 37 per cent of all children born are underweight compared with a national rate of 27.5 per cent. Indicative of household wealth and child welfare is the underweight percentage. 'This therefore provides a strong indicator of the overall welfare situation which is poor and alarming based on international standards' (AIP 2009, Attachments 1–3). It is for these underlying health reasons that the project design document suggested it was advisable to exclude women from peatland development activities.

By taking into account the specific profile of Dayak Ngaju women in the KFCP area, the project document recommended 'a better balance in the design of livelihood interventions'. It was envisaged this would require introducing technological improvements and tools to promote increases in productivity. Consequent increases in productivity would thus improve the bargaining power of women and promote value chains that would be more equitable. Further, it was recommended that due to limiting the access of women in the management of peatland development activities, a payment mechanism would need to compensate for this limitation. As well, the design of the livelihood stream would need to take into account the poor health situation and, for this to be effective, more investment would be needed to target women (AIP 2009, Attachments 1–4).

However, the overarching objective of KFCP was never primarily designed with women in mind or even as a much-needed health care intervention as the project area's overall health profile demonstrates; it was about carbon mitigation. That is, '[t]o demonstrate a credible, equitable, and effective approach to reducing greenhouse gas (GHG) emissions from deforestation and forest degradation, especially from the degradation of peat lands'. Sitting atop the 'KFCP Objective Tree', above the overarching objective, was KFCP's stated goal/impact – 'To inform a post-2012 global climate change agreement and enable Indonesia's meaningful participation in future international carbon markets' (AIP 2009, Figure 2, p. 17).

Reducing GHG emissions, additionality, international carbon markets, MRV, climate change agreements, UNFCCC negotiations, gender and gender mainstreaming are all terms that are 'very high bahasa, bahasa orang [very high language, English people]' (M Central Kalimantan 2014). Not only did one respondent privately concede that KFCP, at least for the participating local communities, was never really about carbon but livelihoods (RD2 Central Kalimantan 2015), but they questioned REDD+ as the appropriate mechanism to reduce emissions from deforestation and forest degradation by asking what the Indonesian government refuses to address. 'Like what is really the problem. Is it really REDD or is it about land use' (RD2 Central Kalimantan 2015). While the former observation brings an acuity to the discontinuity between the project's objectives and goals, and the priorities of the participating Dayak Ngaju, the latter question strikes at the credibility of REDD+ as an effective mitigation mechanism altogether.

The participant's citing of 'land use' is an implicit reference to land concessions, land clearing (burning), logging, expanding oil palm plantations and mining activities, the real drivers of deforestation in Indonesia and other REDD+ countries (Maxton-Lee 2018, 2020; Delabre et al. 2020; Global Forest Coaltion 2020). These are issues, in part, that a sovereign Indonesian government could adopt, if they chose to, through national laws, regulations and carbon taxes (AP1 Jakarta 2014). However, 'reducing the activities that cause greenhouse gas emissions; through strengthening regulations and government oversight … represents a threat to the most profitable sectors of capitalism' (Maxton-Lee 2018, p. 47).

130 *Where are the women?*

As for an 'equitable' approach in the stated KFCP objective above, this falls woefully short when the livelihood stream does not even meet adat women's stated needs – '[they] ask something for health, ask to manage our family in this village and then, until now, KFCP did not give a response for that' (M Central Kalimantan 2014). The word equitable is particularly incongruous on the ground when orang-utans even have their own doctor, while the participating Dayak Ngaju do not.

> It is much more like the feelings of inequality of why they [KFCP] are much more care about the orang-utan than the human ... so like they saw the orang-utans get treated by doctor, they have medical doctors.
>
> (RD2 Central Kalimantan 2014)

Four years after KFCP started and the project officially closed in June 2014, access to, and the lack of health care practitioners prioritised by the women, had not transpired. This, in spite of the project document assessing the overall health and well-being in the area to be 'poor and alarming'. While the project document correctly states that to effectively address the poor health situation, more investment is needed to target women, the policy means through which to achieve this, mainstreaming gender, is as the previous discussion has demonstrated, inadequate. 'Gender mainstreaming in REDD will most likely need to generate a separate funding stream for women controlled and managed activities. These should be located in areas close to the house such as nurseries' (Australia Indonesia Partnership (AIP) PD 1–4 2014).

> So some of them [women] hopefully work for that project to capture the project because they want KFCP project can help them to make a better life, to make maybe give some job to them and help them to give a money for family ... they already talk to them [KFCP] what they want with the meeting ... some woman needed something special job for them they are not giving. KFCP stop just doing their project. Like they already manage the project and just doing that in that village but not what the people wants, especially what woman [wants].
>
> (M'y Central Kalimantan 2014)

There was no discrimination between men and women participating in seedling and planting activities in the KFCP livelihood stream. However, not everyone who wished to participate could.

> You know, project is limited on who will get beneficiaries, it's limited of time, it's limited of budget, so not everyone in village will come into activities. Only some of peoples, some of men, some of women will come because of the limited budget.
>
> (M Central Kalimantan 2014)

What specifically did the adat women express they wanted in the livelihood stream?

> M'y: [N]ot just fresh fish but so they can get more value for that, they needed some skill for to make that. AG: And so, does the livelihood programme provide any of that? M'y: No we have just been doing like other plantations help them to give some seed for rubber plantation and then making seedling for reforestation and then make traditional way to get fish for some people, so just that for the livelihood projects … nothing special for woman. Some of the woman wants they can get a skill for to make make-up and make dress … Special for woman so they can work in home. Going to field, so far from the family.
>
> (M'y Central Kalimantan 2014)

These responses demonstrate how the KFCP project document perpetuates a well-worn 'solution' to women's needs that is emblematic of policies advocating market-based solutions in general that promote 'the idea that income generation programs or women's involvement in commodity/value chains would solve all women's problems, problems in fact that arise from unequal relations of power' (Arora-Jonsson 2014, p. 302). And no clearer are power relations promoted and reflected than through the myth that markets are the solution to forest degradation and deforestation (Delabre et al. 2020).

Carbon markets

The previous chapter's general findings discussed how carbon markets, mitigation projects such as REDD+ and emissions trading are generally viewed with scepticism (PA Central Sulawesi 2014), and caveats of 'NO Rights, NO REDD+' by customary communities (D Central Kalimantan 2014). Another attributed the whole notion of REDD+ as being outside of their 'regime of familiarity' (Thevenot, cited in A2 Jakarta 2014). When participants were questioned specifically about adat women's understanding of carbon markets, responses varied from 'they have no idea' to 'they don't understand – even I don't understand' (D Jakarta 2014).

> Well I'm not saying indigenous women aren't smart enough. Speaking of access, speaking of information, maybe if they have more information but as far as I know if the rules were communicated by REDD in the villages, they don't really understand … they don't understand what REDD is.
>
> (N Jakarta 2014)

> The understanding of the women from the grass-root, they don't have an understanding of that.
>
> (A Jakarta 2014)

132 *Where are the women?*

OK she said in the thinking of the indigenous women, carbon market is more [inaudible] to another country ... she says that because they [adat women] are thinking – Oh my, our peatland is going to another people to another country to bule [foreigner] to international country.

(M Central Kalimantan 2014)

They don't really get it. REDD thing is really a top-down policy that doesn't really come from the people. When they are talking about improving livelihood ... It is not first of all asking them what they want to do, what they want best for their life and livelihoods so this carbon, not carbon market, what carbon is they don't know so can you imagine.

(S Jakarta 2014)

The confusion surrounding notions of carbon/carbon markets/carbon offsets by adat women is similar to that of gender/gender equality/gender mainstreaming; however, incomprehensibility is not the only similarity these two sets of conceptual terms share. Both carbon markets and gender mainstreaming employ technocratic processes to achieve their end results, respectively emission reductions and gender equality. Gender equality, as has been discussed here, is largely equated with 'meaningful participation'. In terms of implementing gender mainstreaming, however, the instruments defined are more precise – data gathering, target setting, monitoring and evaluation (Meier & Celis 2011). Emission reduction instrumentation is also about defining targets, establishing baselines, setting new baseline targets, additionality, monitoring, evaluation and reporting.

What is fundamentally at issue here is a profound disconnect between two vastly different cultural systems: one that privileges traditional knowledge and socio-ecological connectedness; the other steeped in principles of rationality and scientific knowledge in the pursuit of 'progress' presenting as 'ongoing efforts to "fix" and resuscitate failing market based mechanisms' (Delabre et al. 2020, p. 5). When analysed in this light, REDD+'s participatory processes may be viewed 'as sites of knowledge production rather than "add-on" processes' (Delabre et al. 2020, p. 6). On a sub-political level, the construction of this type of procedural inclusion 'is also about corruption and power, and how those with power and privilege dilute, sideline and destroy, block social changes which challenge their power. Where "tick the box" is an example of this' (E Pittaway 2016, pers. comm., 3 May).

Justice

The third and final thematic area the semi-structured questionnaire investigated was justice. Justice is a key concern as evidenced by the overarching question this book seeks to address: 'What is the basis of a just adaptation and mitigation policy once these have been subsumed within a global carbon market framework?'

One aim of the study was to understand how the ubiquitous terms of 'climate justice' and 'gender justice' are understood by the study's participants, and those of the adat women whom they represent. The responses to these terms are then local understandings of justice and are subsequently incorporated into addressing the key question in the following and final chapter: 'Justice in the age of the Anthropocene: reintegration as the fourth dimension of justice and the injustice of maladaptation'.

When exploring the notion of justice in REDD+, participants were invited to reflect on what climate justice and gender justice mean from their own perspective and those of the adat women and communities they represent. One of the participants working in Jakarta attended UNFCCC COP annual talks and had worked closely on REDD+ issues in Indonesia in the sites under study. Climate justice, for this participant, inheres several principles.

> First is the polluter pays principle … but there is also the human security aspect in there. What I understand is that whatever we do to mitigate or adapt to climate change must not undermine our human security, our freedom from fear, freedom from want OK and, also the lands right aspect. This is what I'm talking about with agrarian control.
>
> (AP2 Jakarta 2014)

Climate justice is thus linked to notions of who is historically responsible for anthropogenic climate change and deforestation, that is, 'who pays'; human security and well-being; and agrarian control, where the latter, that is, securing adat tenure, is a pathway to ensuring that human security and its enabling conditions of freedom from want and from fear are maintained. Invariably, participants' responses differed to those of AP2's in their wording of what climate justice means. However, the substance of the broad principles of justice articulated in this response were shared by participants across the three study sites under investigation.

Polluter pays principle (PPP)

While there is no explicit reference to the Polluter-Pays-Principle (PPP)[2] in the UNFCCC, Article 3.1 implicitly recognises this through the fundamental principle of 'common but differentiated responsibilities' (CBDR). That is,

> The Parties should protect the climate system for the benefit of present and future generations of humankind, on the basis of equity and in accordance with their common but differentiated and respective capabilities. Accordingly, the developed country Parties should take the lead in combating climate change and the adverse effects thereof.
>
> (UNFCCC 1992, Article 3.1)

134 *Where are the women?*

The exclusion of the PPP from the UNFCCC, considered by some to be the 'cardinal principle for solving the intractable problem of climate change' (Khan 2015, p. 639), extended to the Kyoto Protocol, which locked into international law a market-based mitigation solution to climate change and was perceived as a triumph for the neoliberal project of market environmentalism (Liverman 2009). The Paris Agreement also does not mention the PPP. Rather than internalise the costs of pollution in production processes in the global North, elaborate schemes such as offsetting mechanisms were constructed, where it only results in moving damage from one place to another. Offsetting, in itself, does not do anything to reduce greenhouse gas emissions or conserve forests or species (McAfee 2014). It is a zero-sum game.

What is evident is the Clean Development Mechanism (CDM) with its inherent flaw of environmental integrity (Cames et al. 2016; see Chapter 3), emissions trading, and cap and trade schemes, where 'what matters is the "cap", but you can't talk about that because that is command and control' (McAfee 2014), and later REDD+, modelled on the CDM and Payment for Environmental Services (PES), now 15 years old, has failed to achieve what it said it was designed to: reduce deforestation (Global Forest Coalition 2020).

The Global Forest Coalition (2020) publication built on the Center for International Forestry Research's (CIFOR) (2018) review of the first 10 years of REDD+ (from 2005 to 2015). The comprehensive failure of this mechanism is well catalogued in the review, and concluded: there was limited meaningful participation with rights holders, lack of respect for indigenous peoples' Free Prior and Informed Consent (FPIC), and overall insufficient attention given to integrating local needs; most REDD+ strategies are gender blind, where the concern for gender issues is not only limited, but prevalent among those working on REDD+ at a national organisational level in the global South; issues of compensation and beneficiaries of REDD+ remain unclear, land tenure and indigenous peoples' and local communities' rights have featured prominently in policy-making at the REDD+ level, but not enough has been done to ensure REDD+ projects are functioning in this regard; REDD+ has failed to deliver improved conservation of biodiversity, other environmental services or economic development and supporting livelihoods; and finally, '[i]nformation and discourses about the drivers of forest change are often purposely hidden or neglected by powerful agents, hindering the transformational changes needed in land-use decision-making' (Angelsen et al. 2018, cited in Global Forest Coalition 2020, p. 2).

A structural critique of REDD+, as with other PES and biodiversity offsets, exposes their dependency on *external* sources of investment or aid growth, for market-based financing (McAfee 2014). The theory is that 'carbon and biodiversity markets will transfer dollars for development to the global South, but incentives to invest in these offsets depend on their profitability and therefore require more economic growth' (McAfee 2014). More economic growth is promoted by companies' need to continually increase their return to shareholders (Maxton-Lee 2020), and on it goes. The major cause of environmental

Where are the women? 135

degradation and deforestation in Indonesia and elsewhere, as Maxton-Lee's (2020) research correctly diagnoses, is not corruption, or lack of stakeholder engagement, or lack of transparency at a country level. 'The crux of the problem is the global economic system in which the only measure of success is increasing short term financial returns. Capitalism, in other words' (Maxton-Lee 2020). Lovelock (2006) may very well have been speaking about capitalism itself when referring to the present state of Gaia, a metaphor for living Earth, in his book the *Revenge of Gaia*, where he points out 'we are currently trapped in a vicious circle of positive feedback' (Lovelock 2006, p. xiii). What started for McAfee's neoliberal natures scholarship as 'Selling nature to save it?' in ecosystem markets a decade and a half ago, has turned to asking 'Can capitalism save itself?' (McAfee 1999, 2014).

REDD+, however, is undeterred by either capitalism in crisis or its consequent climate crisis; its epicentre is 'accumulation by decarbonisation' (Bumpus & Liverman 2008), thus it thrives on both. The neoliberal 'solution' is to continue a well-worn trajectory on retrofitting failed approaches (CDM – see Chapter 4), that 'distract from consideration of alternative and potentially more effective forms of forest governance' (Delabre et al. 2020, p. 5).

Several of the study's participants endorsed alternative approaches to climate change mitigation in their responses, by asking their own questions: 'Why the REDD+ mechanism? Why the emphasis on the carbon-centric MRV to achieving emissions reductions to the exclusion of "community-monitoring"?' (AP2 Jakarta 2014); Why 'create just something that is imagination?' (D Central Kalimantan 2014). The latter participant, part of 'the pragmatists' group, did not reject REDD+ completely, but spoke to a wider-held view about REDD+ in Central Kalimantan. That is, why does the UNFCCC go to the trouble of instituting MRV in REDD+ as the only valid methodology to carbon emissions 'when the adat communities already have evidence that is there' (D Central Kalimantan 2014). In other words, the work of adat communities in protecting and preserving their forests has always gone on 'whether there is REDD or there is not … if there is possible benefits for them, rewards there because they protected area, D said they have rights to get that' (D Kalimantan 2014).

However, proxy indicators to emission reductions or co-benefits such as increasing or decreasing biodiversity, increasing or decreasing forest cover, and current livelihoods of the people are not quantified or monetised like 'measurable' emission reductions in REDD+. When the idea of community monitoring, for example, was first presented to the UNFCCC in 2009,

> the idea was pushed back by the donors mostly, because they wanted a robust carbon emission reduction so they can allocate money to pay for that. And they courted our arguments about the social and economic aspects by putting a label 'non-carbon benefits' which is now actually in the legislation.
>
> (AP3 Jakarta 2014)

136 *Where are the women?*

The problems of REDD+ in Central Kalimantan and its perceived injustice have largely been attributed to this carbon-centric bias to the exclusion of community knowledge systems of sustainable forest management practices. Even though the purpose of REDD+ is acknowledged 'as good, we have to combine the idea of adat community and UNFCCC and then from this idea, when we combine it we will find the way for the solutions. But so far, it's not combining the both side idea but it is just one way, the idea from [donor] Norway' (D Central Kalimantan 2014).

'Community monitoring' did not feature formally in UNFCCC negotiations but had been discussed in side events and lobby sessions with negotiators 'in the hope that it filters into the mainstream discussions' (AP2 Jakarta 2014). For this respondent, its meaning has a dual interpretation. That is, '[f]irst, you work with community in understanding the monitoring and, second you use their methodology, their knowledge, you use their systems' (AP2 Jakarta 2014). For example, in one community forestry in West Sumatra, their understanding of MRV is to make a sampling plot, to box an area of the forest which is then carefully monitored. Leaves, soil and even the leaves from the river are collected. Using this method to monitor biodiversity loss, fewer varieties of these plants have been observed than say 10 years ago. Additionally, patrol teams in West Kalimantan have employed community monitoring, where the adat communities know what plants are on the brink of extinction and, accordingly, do not use these plants. However,

> they [UNFCCC and donors] still want carbon and, communities don't understand [MRV], so they know what plants can be eaten, that can be turned into medicines, that can be used for women delivering birth, but if you ask, OK this one Durian tree, how much carbon is inside this tree, it's very difficult.
> (AP2 Jakarta 2014)

The polluter pays principle, like adat knowledge and methodologies of monitoring biodiversity and forest loss, thus share a common theme. All have been side-lined as valid alternatives in mitigating climate change. In the first case, the PPP would provide a more transparent method to ensure that the cost of deforestation, one of the main drivers of climate change, is borne by those most responsible in Indonesia, for example, logging, mining and palm oil forestry companies (D & N Jakarta 2014). With respect to community monitoring, the devaluation of adat systems and knowledge and, by extension, the adat communities themselves, is largely ignored, except in the 'side events' of the UNFCCC COP meetings, where 'it is still a struggle to put community monitoring in the UNFCCC level' (AP2 Jakarta 2014). This is what climate injustice means for the adat communities, the study's respondents conveyed.

Human security and agrarian control/land rights

Human security and agrarian control, the other two principles of climate justice cited by participant AP2 (Jakarta 2014), though not explicitly referenced as such

Where are the women? 137

by other respondents, were implicit in their responses. The overlapping nature of these principles have resulted in combining both under one heading due to various underlying, unresolved and unaddressed historical and contemporary issues that Indonesian adat communities have experienced and continue to. For example, one of the general findings spoke of the revolving door of international development projects in Central Kalimantan, where the same project developers are 'coming and going under different names, World Bank, Asian Development Bank, Suharto regime, central (Indonesian) government, military; so they [adat communities] say, REDD+ is just one of them' (AP2 Jakarta 2014). Another claimed REDD+ is simply another 'debt problem'.

> I don't see that this REDD, it is just another debt problem. Why? Because it is a South and it doesn't even recognise the problems, the genuine problems, that exist in the society and we are talking about the social structure, indigenous peoples' structure, when they are living in the forests they have already been impacted by the earlier system that you know, sometimes I have difficulty with the words, but you know if they don't recognise the system, how can they solve problems just by giving benefit sharing [under REDD+].
>
> (D Jakarta 2014)

Justice is thus situated within a wider context and background of past and current agrarian injustices. These have been perpetrated by national and international actors impacting on adat livelihoods and their ability to control their own destinies that continue to be reproduced today. One participant who had been researching Borneo's agrarian history at the time of their interview found that as early as the 14th century there had already been internal migration that cost the nomadic Dayak, the original ethnic group, the loss of their forests, resulting in them having to move 'so they could continue to cultivate the rattan again. Because there was ... a Muslim Sultanate there, and later, they interact a lot with the VOC, the Netherlands' (SND Jakarta 2014). In consequence, there has been a 'long history of market penetration ... [which] suddenly turned them all into the commodification of forests. So the commodification of forests equals money' (SND Jakarta 2014). During the course of the participant's research on KFCP, they came across a PowerPoint presentation where project developers presented to participating villagers a 'picture of money REDD+ = $' (SND Jakarta 2014).

The aspiration of adat agrarian control, or self-determination, and livelihood security (AP2 Jakarta 2014), has its antecedents in centuries-long dispossession of the land and resources that sustain their livelihoods. Maxton-Lee's (2018) research into narratives framing conservation efforts of tropical forests in Indonesia, says the early foundations 'were laid down in the colonial period, when the Dutch East Indies government set up nature reserves that were designed to restrict access to the commons by Indigenous people, from this period onwards have been depicted as the main offenders in environmental

138   *Where are the women?*

destruction' (Maxton-Lee 2020, p. 50). The post-Sukarno 'Green Revolution' campaign, under the New Order of the Suharto Regime, was also singled out by the globalist respondents (D & N Jakarta 2014). According to Hansen (1972), in the context of Indonesia's ongoing rice production problem, by mid-1968 the government had signed a contract with Swiss chemical and pharmaceutical firm Ciba, 'to saturate 300,000 hectares of prime rice lands on Java with high yield seeds [miracle rice seeds], fertilizer and pesticides for the 1968–1969 wet season' (Hansen 1972, p. 937). As well as Ciba, a number of other foreign firms from West Germany (Hoeschst, A.H.T.), Japan (Mitsubishi), and a company registered in Europe that had the backing of Indonesian entrepreneurs, Coopa, were contracted to undertake the large-scale programme/campaign to achieve a Green Revolution by 1973 (Hansen 1972). By 1969, ecological dangers were coming to the attention of government authorities due to the widespread programme of aerial spraying. Inland ponds were being poisoned by aerial pesticides. Additionally, massive spraying was having harmful effects on fish and livestock, as well as humans. By May 1970, an official announcement was made by President Suharto 'to abandon the program and terminate the government's relationship with foreign firms' (Hansen 1972, p. 44). Against an historical legacy of colonial rule and decades of ecologically destructive state-sanctioned national and international interventions, REDD+ is simply viewed as a modern-day iteration of past agrarian injustices 'with (carbon) market penetration at its heart' (SND Jakarta 2014).

Responses to the question of what climate justice means for adat women, inevitably followed a meandering pathway through these historical facts to arrive at a simple, yet profound, definition expressed by one participant: 'How can they feed the family, it's what they call justice' (SND Jakarta 2014).

The principles of justice articulated by AP2 (Jakarta 2014) resonated in the mandate of one of the participant feminist organisations whose work centred in the provinces of Central Sulawesi, Central Kalimantan and Ache, to 'strengthen women, grass-root women, in the fight against impoverishment' (A Jakarta 2014). Impoverishment was linked to a range of issues, for example, food sovereignty and conflicts over natural resources among others, where impoverishment is a clear example of how securing adat tenure to ensure freedom from want and from fear, can neither be theoretically, nor practically, separated. Respondent 'A' (Jakarta 2014) was emphatic, reiterating on several occasions throughout the interview that 'we cannot put them in boxes; this is indigenous women and this is local women'. For example, 'in Central Sulawesi, when we are coming into the villages, we are not only seeing the indigenous women. With the transmigration of the land, there is more likely local people now, not originally from the area' (A Jakarta 2014). With respect to REDD+, one aim of this participant feminist NGO is to provide more information to the women outside of the official information they receive from the government and the REDD+ project developers. For example, 'we are also trying to say that there's injustice, global injustice because of the climate change, who's the actor and everything. But, if you want a quick answer, they

Where are the women? 139

[women] don't get it about the climate change justice' (A Jakarta 2014). The explanation for why 'they don't get it', is because the term is elitist, 'it is too far away from the ground, what is climate change?' (A Jakarta 2014). This response mirrored that of another in Central Kalimantan, where the participant explained the difficulty of translating what climate change mitigation is. Firstly, it requires an understanding of what climate change is; which in turn requires explaining the mechanisms, such as REDD, meant to reduce carbon emissions. And then, 'like what is carbon, what is emission, how to reduce it and that kind of thing' (RD Central Kalimantan 2014). The use of these 'high terminologies' (RD Central Kalimantan 2014) made one respondent question,

> Well why do they plan to use those abstract terms [in REDD+] that are so far from daily life of women on the ground? Why do they choose to do that? Why do they try to sanitise it from things relevant to a woman's life, children, water, fishing, so this at the UNFCCC level, but of course at the UNFCCC level it is very abstract, very far from the ground, but why should it be like that?
>
> (AP2 Central Kalimantan 2014)

While climate justice, and gender justice, has led to no single definition in the participant responses, they are relationally implied in several principles underscoring 'climate justice' by AP2 (Central Kalimantan 2014). Namely, the polluter pays principle, human security and agrarian control, the land rights aspect. These principles are linked to historical facts framing current claims for adat tenure; recognition of alternative methodologies that respect adat cultural systems, local knowledge, practices and expertise side-lined in the mainstream discussions of the UNFCCC as strategically inferior to carbon centrism; the prevalence of relations of power and the asymmetry of power manifest in abstract terms, that have little meaning to those on the ground – that is, participating customary communities in REDD+.

Returning to notions of gender justice and climate justice by way of conclusion, participants were asked about their understanding of the phrase, 'there can be no climate justice without gender justice' (Gender 2007, cited in Rooke 2009). This produced a similar pattern of responses to those regarding notions of what climate justice and gender justice means; the logic of response was often contextual, tangential and, in this case, inverted. That is, in defining what this expression of justice meant, injustice was first made visible.

> So what I think about there can be no climate justice without gender justice is we have to make the layers of injustice visible first. And, when we talk about actions, efforts to mitigate climate change, we have to really scrutinise those mechanisms, REDD+. Does it avert inclusivity of women or, it just operates on the status quo principle ... the assumption that it [climate change] is gender neutral.
>
> (A2 Jakarta 2014)

140 *Where are the women?*

The assumption climate change is a gender-neutral issue was first introduced in Chapter 1 and results from climate change being defined, first and foremost, as a science problem, hence the focus on mitigation. This scientific assumption of gender neutrality has been accompanied with another, from a different episteme, in the sphere of the 'market', where feminists from the neoclassical school view the 'market as sexually neutral' (Bessis 2003, p. 1). Epistemic assumptions from both have been structurally debunked in presenting these empirical insights from the women-specific findings of the Indonesian study. No amount of fine tuning of REDD+, however, can remedy the deep androcentrism at its heart.

Conclusion

'Where are the women?' is a question anticipating their ongoing exclusion; a questioning point of priority and reflection for decision makers and stakeholders. It is unsurprising CIFOR's review of the first 10 years of REDD+ found it gender blind, given the systemic disinterest about gender issues in general in implementing countries of the global South (Angelsen et al. 2018, cited in Global Forest Coalition 2020). It is unsurprising for two reasons: firstly, fish rots from the androcentric head down; secondly, that is because the status of nature, women and customary communities in REDD+ are metaphorically buried beneath layer upon layer of oppression depicted in the status quo 'Iceberg' (Bennholdt-Thomsen & Mies 1999) and 'Layer Cake with Icing' (Henderson 1982) models of global neoliberal economy. Gender mainstreaming and other participatory processes, procedural inclusion, to achieve asexual market equality illuminates their neoliberal pedigree as 'silos of knowledge production' (Delabre et al. 2020). Participation at the local level cannot be isolated from a wider political project of empowerment; to do so continues perpetuating 'historical and current exclusion and leading to erasures of complexities and diverse ways of knowing' (Delabre et al. 2020, p. 6).

Notes

1 A similar conflation with masculinity and markets was drawn by an Australian female participant well over two decades ago at the Conference of the Parties (COP) 3 in Kyoto in 1997 (Gender CC). 'COP 3 Kyoto 1997: Missing women's organisations. Records suggest that there were some women's activities and position papers distributed at COP 3 in Kyoto, although these are no longer available. Indeed, there was an email send [sic] out by an Australian participant asking why women's organisations were not more strongly present at COP 3 and why they did not take more notice [of] the conference. Partly answering the question, herself, she wrote: "The arguments used here are almost entirely economic. Decisions are made mostly with little consideration being given to survival. Perhaps women felt they could not penetrate this masculine perspective – and stayed at home"' (Gender CC, www.gendercc.net/genderunfccc/unfccc-conferences/kyoto-1997-to-new-delhi-2002.html). This speaks to the androcentric nature at the heart of international organisations of which Bessis (2003) wrote in her sweeping review of women and UN agencies (see Chapter 5 of this book).

2 'The principle to be used for allocating costs of pollution prevention and control measures to encourage rational use of scarce environmental resources and to avoid distortions in international trade and investment is the so-called "Polluter-Pays-Principle". This principle means that the polluter should bear the expenses of carrying out the above mentioned measures decided by the public authorities to ensure that the environment is in an acceptable state. In other words, the cost of these measures should be reflected in the cost of goods and services that cause pollution in production and/or consumption. Such measures should not be accompanied by subsidies that would create significant distortions in international trade and investment' (Khan 2015, p. 640).

References

Alston, M 2014, 'Gender mainstreaming and climate change', *Women's Studies International Forum*, vol. 47, pp. 287–294.

Arora-Jonsson, S 2014, 'Forty years of gender research and environmental policy: where do we stand?', *Women's Studies International Forum*, vol. 47, pp. 295–308.

Australia Indonesia Partnership (AIP) 2009, *Kalimantan forests and climate partnership (KFCP) design document*, Indonesia-Australia Forest Carbon Partnership, accessed 15 January 2015, http://formin.finland.fi/public/download.aspx?ID=48885&GUID=%7B9B0BA3BA-25BF-4FEA-985B-B6DADCA60EAC%7D.

Baden, S & Goetz, A 1998, 'Who needs [sex] when you can have [gender]', in C Jackson & R Pearson (eds), *Feminist visions of development: gender analysis and policy*, e-book, accessed 10 June 2017, books.google.com.au, pp. 19–38.

Barnett, J & O'Neill, S 2010, 'Maladaptation', *Global Environmental Change*, vol. 20, pp. 211–213.

Bennholdt-Thomsen, V & Mies, M 1999, *The subsistence perspective: beyond the globalised economy*, Zed Books, London UK & New York, USA.

Bessis, S 2003, 'International organisations and gender: new paradigms and old habits', *Signs: Journal of Women in Culture and Society*, vol. 29, no. 2, pp. 633–647.

Boyd, E, Hultman, N, Roberts, J, Corbera, E, Cole, J, Bozmoski, A, Ebeling, J, Tippman, R, Mann, P, Brown, K & Liverman, D 2009, 'Reforming the CDM for sustainable development: lessons learned and policy futures', *Environmental Science & Policy*, vol. 12, no. 7, pp. 820–831.

Bumpus, A & Liverman, D 2008, 'Accumulation by decarbonisation and the governance of carbon offsets', *Economic Geography*, vol. 84, no. 2, pp. 127–155.

Cames, M, Harthan, RO, Füssler, J, Lazarus, M, Lee, CM, Erickson, P & Spalding-Fecher, R 2016, *How additional is the Clean Development Mechanism?: analysis of the application of current tools and proposed alternatives*, Study prepared for DG CLIMA, Öko Institut, accessed 7 August 2020, https://ec.europa.eu/clima/sites/clima/files/ets/docs/clean_dev_mechanism_en.pdf.

Delabre, I, Boyd, E, Brockhaus, M, Carton, W, Krause, T, Newell, P, Wong, GY & Zelli, F 2020, 'Unearthing the myths of global sustainable forest governance', *Global Sustainability*, vol. 3, no. 16, pp. 1–10.

Global Forest Coalition 2020, '15 years of REDD+: has it been worth the money?', *Global Forest Coalition*, accessed 4 September 2020, https://globalforestcoalition.org/wp-content/uploads/2020/09/REDD-briefing.pdf.

Hansen, G 1972, 'Indonesia's green revolution: the abandonment of a non-market strategy toward change', *Asian Survey*, vol. 12, no. 11, pp. 932–946.

142 *Where are the women?*

Henderson, H 1982, *Total productive system of an industrial society: layer cake with icing*, digital image, accessed 14 May 2017, http://hazelhenderson.com/wp-content/up loads/totalProductiveSystemIndustrialSociety.jpg.

Khan, M 2015, 'Polluter-pays-principle: the cardinal instrument for addressing climate change', *Laws*, vol. 4, no. 3, pp. 638–653.

Lang, C 2009, 'REDD: CO2lonialism of forests', *REDD-Monitor*, weblog post, 16 April, accessed 20 June 2017, www.redd-monitor.org/2009/04/16/redd-co2lonialism -of-forests/.

Liverman, D 2009, 'Conventions of climate change: constructions of danger and the dispossession of the atmosphere', *Journal of Historical Geography*, vol. 35, no. 2, pp. 279–296.

Lovelock, J 2006, *The revenge of Gaia: why the earth is fighting back and how we can still save humanity*, Allen Lane, Camberwell. Foreword by Crispin Tickell.

MacGregor, S 2009, 'A stranger silence still: the need for feminist social research on climate change', *The Sociological Review*, vol. 57, no. 2s, pp. 124–140.

Maxton-Lee, B 2018, 'Narratives of sustainability: a lesson from Indonesia: global institutions are seeking to shape an understanding of sustainability that undermines its challenge to their world view', *Soundings: A journal of politics and culture*, no. 70, Winter, pp. 45–57.

Maxton-Lee, B 2020, 'Forests, carbon markets, and capitalism: how deforestation in Indonesia became a geo-political hornet's nest', Guest Post, *REDD-Monitor*, accessed 22 August 2020, https://redd-monitor.org/2020/08/21/guest-post-forests-ca rbon-markets-and-capitalism-how-deforestation-in-indonesia-became-a-geo-political-hornets-nest/.

McAfee, K 1999, 'Selling nature to save it? biodiversity and green developmentalism', *Environment and Planning D: Society and Space*, vol. 17, pp. 133–154.

McAfee, K 2011, *Nature in the market-world: social and developmental consequences and alternatives*, Draft paper, UNRISD Conference, Green Economy and Sustainable Development: Bringing Back the Social Dimension, 10–11 October, Geneva, United Nations Research Institute for Social Development.

McAfee, K 2012, 'The contradictory logic of global ecosystem services markets', *Development and Change*, vol. 43, no. 1, pp. 105–131.

McAfee, K 2014, *Green economy or buen vivir: can capitalism save itself*, Lunchtime Colloquium, Rachel Carson Centre for Environment and Society, online video, accessed 17 May 2020, www.carsoncenter.unimuenchen.de/events_conf_seminars/ event_history/2014-events/2014_lc/index.html.

Meier, P & Celis, K 2011, 'Sowing the seeds of its own failure: implementing the concept of gender mainstreaming', *Social Politics: International Studies in Gender, State & Society*, vol. 18, no. 4, pp. 469–489.

Mellor, M 2012, *Just banking conference*, Friends of the Earth Scotland, online video, accessed 14 May 2017, www.youtube.com/watch?v=i-9vMXfTM10.

Rooke, A 2009, 'Doubling the damage: World Bank climate investment funds undermine climate and gender justice', *Gender Action*, accessed 10 May 2017, www.gen deraction.org/images/2009.02_Doubling%20Damage_AR.pdf.

Sarantakos, S 1998, *Social Research*, 2nd edn, Macmillan Education Australia Pty Ltd, South Yarra, Australia.

The Forests Dialogue 2012, *Background paper for the scoping dialogue on the exclusion and inclusion of women in the forestry sector*, Women Organising for Climate Change in Agriculture and NRM, accessed 10 June 2017, http://theforestsdialogue.

org/publication/background-paper-scoping-dialogue-exclusion-and-inclusion-women-forestry-sector.

United Nations Programme on Reducing Emissions from Deforestation and Forest Degradation (UN-REDD Programme) 2011, *The business case for mainstreaming gender in REDD+*, The United Nations Collaborative Programme on Reducing Emissions from Deforestation and Forest Degradation in Developing Countries, accessed 15 May 2017, www.undp.org/content/dam/undp/library/gender/Gender%20 and%20Environment/Low_Res_Bus_Case_Mainstreaming%20Gender_REDD+.pdf.

United Nations Programme on Reducing Emissions from Deforestation and Forest Degradation (UN-REDD Programme) Indonesia 2012, *Central Sulawesi's readiness to implement REDD+ after 2012*, accessed 20 June 2014, www.unredd.net/docum ents/un-redd-partner-countries-181/asia-the-pacific-333/a-p-partner-countries/indonesi a-187/communication-knowledge-sharing-1545/others-1677/7226-central-sulawesi-read iness-to-implement-redd-2012-7226.html.

United Nations Framework Convention on Climate Change (UNFCCC) 1992, accessed 9 May 2017, https://unfccc.int/resource/docs/convkp/conveng.pdf.

Women Organizing for Change in Agriculture and Natural Resource Management (WOCAN) 2012, *UN-REDD Program: A guidance note to integrate gender in implementing REDD+ social safeguards in Indonesia*, Women Organizing for Change in Agriculture and Natural Resource Management, accessed 11 October 2014, www.wocan.org/resources/guidance-note-integrate-gender-implementing-redd-s ocial-safeguards-indonesia.

Zalewski, M 2010, '"I don't even know what gender is": a discussion of the connections between gender, gender mainstreaming and feminist theory', *Review of International Studies*, vol. 36, no. 1, pp. 3–27.

8 Justice in the age of the Anthropocene

Reintegration as the fourth dimension of justice and the injustice of maladaptation

Introduction

Guiding this concluding chapter, several aims structure the discussion. Firstly, to provide some background thoughts on justice and reassess critical theorist Nancy Fraser's (2008) theory of justice in light of findings of the Indonesian study and the general arguments that have emerged in this book. As Fraser's three-dimensional justice framework stands (with its corresponding injustices) – economic redistribution (maldistribution), cultural recognition (misrecognition) and political representation (misrepresentation) – it is insufficient to accommodate the specific injustice of maladaptation whose remedy is found in environmental 'reintegration'. Hence, there is a need to reassess the scope of the tripartite framework to include this fourth dimension (reintegration/maladaptation), thus preserving the alliterative framework which characterises Fraserian justice.

While no claim is made here to theorise justice dialogically, that is, 'through the give-and-take of argument through democratic deliberation' (Fraser 2008, p. 167) on the depth and scale Fraser has dedicated in developing her theory of justice – a project that has unfolded over two decades – the aim here is more modest. It is a preliminary introduction and discussion to outline the contours of this fourth dimension of justice, reintegration and, in particular, its corresponding injustice of maladaptation, which have profound implications for mitigation and adaptation policies that are more effective in achieving justice beyond rhetoric and thus avoiding the injustice of maladaptation.

Firstly, Fraser's tripartite framework is employed as a conceptual intervention to overlay a catalogue of injustices, revealed through conceptual research undertaken in earlier chapters, such as women as 'silent offsets', as well as metaphorical conceptualisations such as the 'Layered Cake with Icing' (Henderson 1982), 'Iceberg Model' (Bennholdt-Thomsen & Mies 1999), and Mellor's (2012) 'Economic Dualism'. These models are representative of the structure of contemporary patriarchal capitalist economies. A similar undertaking with findings from the Indonesian study follows. In particular, the major finding of recognition of adat tenure and its ongoing jurisdictional implementation by the Indonesian state government.

Justice in the age of the Anthropocene 145

Secondly, the discussion returns to migration and displacement resulting from maladaptive climate change solutions, that is adaptation and mitigation policies under the United Nations Framework Convention on Climate Change (UNFCCC). Maladaptation is a qualitatively different driver of environmentally induced migration distinct from climate change itself. Accumulating empirical experiences on the implementation of climate change adaptation has seen the area of scholarship on maladaptation increasing in response to assessing and understanding negative consequences from adaptation policies, in particular (Barnett & O'Neill 2010; Juhola et al. 2016; Magnan et al. 2016). More recently, Vigil's (2018) research implicates mitigation as well, in relation to the interactions between climate change policies (biofuels and forest carbon projects), land grabs and displacement, suggesting that scholars and practitioners engaged in environmental migration and displacement need to 'broaden the spectrum of their analyses' to include 'green grabbing induced displacement' (Vigil 2018, p. 370). In general, the author suggests (and this book agrees) that 'there is a need to move beyond the category of environmentally induced displacement in order to include the impacts of climate change mitigation policies as a factor that influences displacement outcomes or migratory decisions' (Vigil 2015, p. 45). Work et al.'s (2019) empirical research in Cambodia on conservation, irrigation and reafforestation projects, highlights the ways in which climate change mitigation and adaptation ('CCMA') initiatives and industrial development are intertwined, 'and this confluence is creating potentials for maladaptive outcomes' (Work et al. 2019, p. 47), concluding, '[w]e face today new dangers from climate change projects and policies as much as we do from the effects of climate change itself' (Work et al. 2019, p. 50).

Thirdly, and finally, the discussion returns to the key question framing this book. That is, 'what is the basis of just adaptation and mitigation policies that can be adopted by the UNFCCC once these have been subsumed within a global carbon market framework?' Fraser's theory of justice cannot be delivered in practice while the current and conventional faith in neoliberal environmental solutions prevail. That is, a faith which subscribes to 'market-driven logics as an accepted methodology for dealing with climate change' (Bailey & Wilson 2009, p. 2335). Indeed '[a] number of commentators have [even] questioned the possibility of reconciling the pursuit of justice with neoliberal environmental governance' (Lohmann 2008; Okereke 2008, cited in Mathur et al. 2014, p. 44). The commodification of carbon by innovative market-based mechanisms adopted by the Kyoto Protocol and retrofitted to the Paris Agreement's new mitigation mechanism under Article 6.4 (see Chapter 3), has failed to address the one baseline that ultimately matters. That is, 'it is the baseline emissions path [business as usual] that must be altered if the problem of global warming is to be resolved' (Wara 2007, p. 596).

The dominant neoliberal environmental and economic approach to forest governance, manifest in the United Nations Programme on Reducing Emissions from Deforestation and Forest Degradation (UN-REDD+) and private

146 *Justice in the age of the Anthropocene*

forest projects, is sustained by a set of myths resulting in associated lock-ins which the authors term 'an ideational crisis' (Delabre et al. 2020, p. 1). For example, 'states manage forests independently for the social benefit' (institutional/legal lock-in); 'sustainability is threatened by small-scale farmers' (social lock-in); 'markets are the solution to deforestation and forest degradation' (economic lock-in); 'what is counted – through valuation – counts' (political lock-in); and 'sustainable forest governance initiatives currently "include" local communities in decision-making' (social/procedural lock-in) (Delabre et al. 2020, p. 7). Each of these myths, and their corresponding lock-ins, singularly and together inhere maladaptation in their construction, by ignoring alternative pathways to more just adaptation and mitigation policies presented in this chapter. For example, the inclusion of alternative customary communities' methodologies and cultural practices, knowledge and systems for managing forests. The Paris Climate Change Agreement acknowledges, in the context of adaptation under Article 7.5, that adaptation action should 'be based on and guided by the best available science, and as appropriate, traditional knowledge, knowledge of indigenous peoples and local knowledge systems' (UNFCCC Paris Agreement 2015). However, similar to women in REDD+, acknowledging indigenous peoples' local systems does not mean you have 'mainstreamed them into the heart of policies' (AP2 Jakarta 2014).

(1) Background thoughts on justice

The following pages present an overview of critical political theorist and philosopher Nancy Fraser's (2008) three-dimensional theory of justice, and latterly a reassessment of this theory in light of a fourth dimension proposed here, 'environmental reintegration' and its corresponding injustice of maladaptation.

In 2009, when first beginning research on climate change policies and their potential negative impact on future migrations, foremost in this author's thinking at the time was an overriding normative question propelling its thrust. What does 'justice' mean once the atmospheric space has been gerrymandered by developed countries to the detriment of developing countries? Conventional political theories of liberalism and communitarian theories of citizenship hold that justice is primarily owed to its citizens. In challenging liberalism's philosophical problem of a closed society, the cosmopolitan critique offers the notion of ethical universalism. This view affirms a liberal impartiality that 'our' rights matter only if the rights of foreign humans matter too (Sen 1996, cited in Ginty 2003, p. 43). On this level, 'all human beings matter and matter equally' (Gibney 2000). Cosmopolitanism in debates about justice then, is the most obvious way in which reflection on the subject of justice outside the citizenry of nation-states is unveiled (Barnett 2010).

However, justice in general, viewed in foundational terms about what ethical obligations are owed to citizens and 'others' is still the focus of such political theories on citizenship and, more importantly, is the emphasis on making these theories as normatively robust as possible. Within this context, it is clear what Baier (1991) meant in asking the question, '"Yes, but am I under a moral obligation to her?" is a mark of intellectual immaturity to [even] raise that question' (Baier 1991, cited in Rorty 1998, p. 185). Rorty (1998) contends, however, 'that we shall go on asking that question as long as we agree ... that it is our ability to *know* that makes us human' (Rorty 1998, p. 185). This is the essence of foundational thinking, knowable truths about who we are and that which makes us human. By extension, such foundational thinking is evident in the privileging of neoliberal market environmentalism as *the* solution to climate change. In the absence of considering alternatives, it then becomes a sign of intellectual dogmatism, or what Delabre et al. (2020, p. 5) describe as 'cognitive lock-in'.

With respect to climate change debates and negotiations, justice as well as equity have featured prominently since the UNFCCC was first adopted in Rio in 1992 (Mathur et al. 2014) – particularly in the context of mitigation and adaptation responses to cope with committed climate changes. The reason these normative concepts have been paramount is related to the deep historical injustice in which global warming has occurred – where developed countries have prospered industrially at the expense of developing countries through fossil-fuelled overconsumption (Adger et al. 2006).

In 1997, the UNFCCC Kyoto Protocol recognised developed countries (Annex 1 countries) as 'principally responsible' for the build-up of greenhouse gas (GHG) emissions in the atmosphere over the past 150 years, thus placing a heavier burden on developed countries to act first in mitigation activities under the principle of 'common but differentiated responsibilities', which Elliott (2007) has described as a form of burden sharing. Who pays for adaptation was agreed to by signatories of the 2009 Copenhagen Accord – that is, 'developed countries shall provide adequate, predictable and sustainable resources, as well as technology and capacity-building support to support the implementation of adaptation action in developing countries' (UNFCCC, Newsletter March 2010 – Adaptation). In summary, the 'what' (anthropogenic GHG emissions) and 'who' is historically responsible for climate change (developed countries), who pays (developed countries) and by how much (developed countries), may be viewed as key questions of justice within a distributional paradigm.

Fraserian three-dimensional justice – redistribution, recognition and representation

In contrast to conventional theories of justice, liberalism and communitarianism, and their concern with robust normativity, Fraser's three-dimensional justice framework is a 'bottom-up' way of theorising justice as not being 'one way or the other' (Barnett 2010, p. 246). The tripartite justice framework is well suited to understand the *scope* of justice claims-making associated with climate change solutions presented here, both conceptually and empirically. Barnett (2010) and Schlosberg (2004) argue theorising justice should not proceed along purely foundational lines such as the 'distributional paradigm'

148 *Justice in the age of the Anthropocene*

(or redistribution) central to liberal theories of justice. In developing the third dimension of social justice, the political dimension of representation, Fraser (2005) argued that 'these theories have so far failed to develop conceptual resources for reflecting on the meta-issue of the frame' (Fraser 2005, pp. 72–73), discussed later in the chapter.

In other words, non-foundational ways of theorising justice, such as Fraser's framework, are more interested in 'attaining *justice*, rather than attaining a sound *theory of justice* ... [hence, the] need to focus on the reasons and processes behind determining [injustices] or the lack thereof' (Schlosberg 2004, p. 520). Fraser's method of problematising and thematising justice is a welcome departure for freeing such concepts of justice from the 'imaginary geographical constraints' (Barnett 2010) of the traditional nation-state frame. Within this frame, justice is debated and disputed, overwhelmingly, in terms of membership and exclusion, duties and obligations, who owes what to whom and why, which is a grammar of political claims-making on justice (Fraser 2008); a grammar at odds in the age of the Anthropocene.[1] Although no reference is made to this paleontological term in Fraser's project, she does engage with the issue of global warming as but one of a range of transnational forces and issues in a globalised age that trespass territorial borders, thus giving rise to injustices outside the 'territorial imaginary' (Fraser 2008). For example, polluting the atmosphere with excessive greenhouse gas emissions that breach the biophysical limits of the planet to the detriment of all living creatures, human beings included. The extra-territorial nature of climate change thus lies in its complete disregard for territorial integrity.

The most general meaning of justice for Fraser (2005, 2008) is parity of participation. That is, 'how fair or unfair are the terms of interaction that are institutionalised in the society? ... In my view, then, justice pertains *by definition* to social structures and institutional frameworks' (Fraser 2004, cited in Dahl et al. 2004, p. 378). These structures and frameworks relate to the economic, cultural and political spheres of society that present obstacles to justice thus leading to distinct kinds or 'species' of injustice. For example, economic injustice is created by class stratification where the economic structure can deny some people the resources that they need to be able to fully participate as peers in society, thus giving rise to maldistribution or redistributive injustice. Cultural injustice (misrecognition) is embedded in internal social hierarchies, the status order, where a person's culture or gender is undervalued, which then prevents them from interacting and participating on par with others, in this case they suffer from social inequality or misrecognition. The third dimension relates to the political structure. On this dimension, political injustice (misrepresentation) emerges from governance structures and decision-making procedures, not only internally but beyond the national frame (Fraser 2008). Through a process of elaboration, amplification and elucidation, or non-foundational ways of theorising normative concepts such as justice in relationship to academic analysis (Brandom 1994, Cooke 2006, cited in Barnett 2010), Fraser's project has made explicit what injustice means, thus enriching the pool of justifications for claimants, and their advocates, in demanding justice on these three dimensions. That is, economic redistribution, cultural recognition and political representation.

Beginning in the 1980s, Fraser argued that welfare state politics should not be framed in distributional terms only, that is, 'who gets what', but also, 'who gets to *interpret* what people need' (Fraser, cited in Dahl et al. 2004, p. 375), thus implicating the cultural dimension. The intersection of these two questions marked the start of theorising social justice as redistribution and recognition before those terms were even coined or, in Fraser's words, '*avant la lettre*' (Fraser, cited in Dahl et al. 2004, p. 375). However, it was not until 2004 that Fraser introduced the third dimension to her justice framework, that is, representation, which problematised procedures of decision-making and governance structures (the jurisdictional frame), pertaining to justice claims-making (Fraser 2004, cited in Dahl et al. 2004; Fraser 2008). Specifically, what the appropriate frame is in which claims for justice should be made in a globalised world, were brought to the fore, where the 'Keynesian-Westphalian frame'[2] that had dominated disputes over justice in the democratic welfare state after World War II up until the 1970s (Fraser 2008) could no longer be taken for granted. That is,

> Distribution and recognition could appear to constitute the sole dimensions of justice only so long as the Keynesian-Westphalian frame was taken for granted. Once the question of the frame becomes the subject to contestation, the effect is to make visible a third dimension of justice [political] which was neglected in my previous work – as well as the work of many other philosophers.
>
> (Fraser 2008, p. 17)

In other words, the meta-issue of the frame, the jurisdictional aspect of political representation, has had the effect of 'misframing justice' (Fraser 2008), thus giving rise to the political injustice of misrepresentation that can be reduced neither to redistribution nor recognition. Democratic representation facilitates the laying out of the ground rules for accessing recognition and redistribution. Fraser (2008) acknowledges that remedying this meta-level issue of the frame is a fraught project. Indeed, as others have noted, 'the choice of the scale is not only important for the analysis of justice – it is also invoked by different actors to strategically construct injustice' (Mathur et al. 2014, p. 43). However, Fraser does offer the 'spatially expansive principle of affectedness, in terms of the "all-subjected" principle, one which trumps membership as a criterion of democratic inclusion' (Barnett 2010, p. 251).

The following section centres on the application of Fraser's justice framework to some of the key conceptual insights and empirical findings in this book, as well as a reassessment of the framework to accommodate the 'fourth dimension' of justice – environmental reintegration and its corresponding injustice of maladaptation.

150 *Justice in the age of the Anthropocene*

A *discussion on the application of Fraserian justice to conceptual and* empirical findings of the Indonesian study: an introduction to the fourth dimension of justice/injustice – environmental reintegration/maladaptation

'Silent offsets'

The scientific framing of climate change as gender neutral by the Intergovernmental Panel on Climate Change (IPCC) has profoundly shaped the way in which women's inclusion in adaptation and mitigation interventions has unfolded. In earlier chapters the notion of women as 'silent offsets' was introduced, that focused on the early silence surrounding women in mitigation and adaptation and, more generally, their exclusion in climate change texts and negotiations, from the UNFCCC Secretariat to the Conference of Parties (COP) (Röhr 2006; Carvajal-Escobar et al. 2008). Later, feminist analysis from Chapter 7, 'Where are the women?', reveals the blunt instrument of gender mainstreaming has devolved from its transformative vision of gender equality (Zalewski 2010) over a quarter of a century ago at the Beijing Women's Conference in 1995, to bland procedural participation via quotas in REDD+. As the participative process in the Indonesian study revealed, and respondents confirmed: 'they just put on the paper only' (S Jakarta 2014); 'it's actually just to speed up the process and present a report to ... the donors and international world' (N Jakarta 2014). This is hardly what Fraser (2008) envisaged as 'parity of participation', her most general meaning of justice, under the framework. The undervaluing of women's status in REDD+ projects is not simply a case of gender injustice, it prevents them from participating on par with others and this leads to the injustice of misrecognition or social inequality.

'Adding insult to injury', to use Fraser's term (in Fraser & Olson 2008), the discredited notion of 'meaningful participation' in gender mainstreaming leading to misrecognition, extends to that of rights holders in REDD+ as the Center for International Forestry Research's (CIFOR) review of the programme's first 10 years found (Angelsen 2018, cited in Global Forest Coalition 2020), concluding that meaningful participation in projects is often limited, disrespectful of the principle of Free Prior and Informed Consent (FPIC) of indigenous peoples, with an overall insufficiency paid to integrating local-level needs. CIFOR's conclusion, in addition to its other systemic findings of REDD+, is that the programme is 'gender blind', which fits the Indonesian study's empirical findings hand in glove: inviting the women to express their needs, and then ignoring them. 'Something for health', is but one cruel example, as the Kalimantan Forest Climate Partnership (KFCP) study revealed.

This finding, of course, implicates the economic aspect of benefit sharing, or equity, under REDD+, which CIFOR found remains unclear as to the nature, level of compensation or even who the exact beneficiaries are (Angelsen et al. 2018, cited in Global Forest Coalition 2020, p. 2). On this level, not only are adat women and customary communities misrecognised, this injustice is compounded by that of another, that of economic inequality inherent in the

Justice in the age of the Anthropocene 151

scheme, giving rise to the injustice of maldistribution. The question of 'who' is (or who ought be) responsible for remedying these injustices is the subject of Fraser's misframing justice: Is it the Indonesian government? Norway in their bilateral donor relationship with the Indonesian government? Australia in the now completed bilateral KFCP with the Indonesian government? Transnational corporations in receipt of REDD+ emission reductions to continue polluting elsewhere in what is well known as a zero-sum game?

The concern here, and the primary consideration which led Fraser (2008) to incorporate political representation as the third dimension of justice, 'is with meta-level political injustices, which arise as a result of the division of political space into bounded polities' (Fraser 2008, p. 146). REDD+ is an example of a meta-level political injustice, because it demonstrates 'the way in which the international system of (supposedly) sovereign states [bounded polities] gerrymanders the political space at the expense of the global poor' (Fraser 2008, p. 146). Applied directly to conditions of carbon commodification in general, and REDD+, the Clean Development Mechanism (CDM), and potentially its Paris successor, whose environmental integrity will be fatally compromised if Kyoto-era carryover credits are permitted to be included as part of a country's Nationally Determined Contributions going forward under the Paris Agreement (Evans & Gabbatiss 2019a, 2019b), the atmospheric space has been gerrymandered by developed countries at the expense of developing countries under conditions of carbon capitalism.

'Economic Dualism' and the gendered silent offset economy

The notion of silent offsets segued into the gendered silent offset economy following a detailed theoretical analysis and discussion in Chapter 5 – *social reproduction is leveraged in the interests of carbon capitalism, giving rise to the gendered silent offset economy*. More broadly the critical gender analysis of climate change adaptation and mitigation policies under the UNFCCC has drawn on intra-disciplinary feminist theories positioned at either end of the political spectrum. Classical liberal feminism assumes that the 'market' is sexually neutral (Bessis 2003), while ecofeminism challenges this assumption through its use of conceptual metaphors for (patriarchal) industrialised economies, such as the Iceberg Model and Layered Cake and, latterly, and in particular, the model of Mellor's (2012) Economic Dualism. In their application to carbon mitigation projects such as REDD+, these metaphors and Mellor's model provide powerful visual explanations for women's silence and invisibility in their application to climate change mitigation and adaptation solutions. What is valued is commodified markets to address the issue of climate change; and what is undervalued, broadly manifest in the institutions of social reproduction, which is the basis of climate change interventions from a human-ecological point of view. These models illustrate 'how the so-called productive economic activities depend on a set of currently invisible processes' (Cameron & Gibson-Graham 2003, p. 8).

152 *Justice in the age of the Anthropocene*

Thus, the assumptions of climate change as gender neutral as well as the sexual neutrality of 'markets' by classical liberal feminists provide powerful explanations for the invisibility and silence surrounding women of the global South affected by climate change mitigation and adaptation policies. In other words, arriving at a definition of what women as silent offsets and the gendered silent offset economy mean has emerged in relation to, and is a synthesis of, these various critiques of climate change solutions. How, then, does Fraser's theory of justice interpret the notion of silent offsets?

The emphasis on 'economy' provided in these metaphors and Mellor's model (2012) appears to identify a maldistributive economic system as the overwhelming claim for redistributive justice. However, that would be an obvious and overly simplistic surface reading. This is not intended to minimise claims for redistribution that such a one-sided economy inherent in the 'Economic Dualism' produces; in the context of responses from participants in the Indonesian study, Schlosberg (2004) is right to claim that 'one cannot talk of one aspect of justice without it leading to another' (Schlosberg 2004, p. 527). Thus claims for justice are multiple and overlapping as the following demonstrates.

In an attempt to understand what climate justice and gender justice meant to adat women, respondents from Jakarta, to Palankaraya and Palu, would contextualise their responses within an historical framework of past agrarian injustices linked to contemporary iterations. For example, the Green Revolution, Ex-Mega Rice Project and current, ongoing environmental injustices, such as oil palm plantations, mining and logging concessions, the imposition of multiple World Bank and foreign-sponsored forest conservation and biodiversity project interventions. To these may be added the Indonesian Environment and Forest Minister's recent announcement that the Indonesian government, as part of its commitment to reducing greenhouse gas emissions, is to speed up efforts to issue a regulation on carbon trading. The announcement follows year-long preparations of 'the emission-trading system for the future domestic carbon market ... the set of regulations would stipulate at least three main issues including guidelines and plans on carbon trading, carbon offsets and the commodity market' (Siti 2020, cited in Oktavianti 2020). The minister 'revealed that movement on the regulation was made after the government confirmed receipt of a US$56 million grant from Norway as the first payment for Indonesia's success in reducing carbon emissions under ... REDD' (Oktavianti 2020). Another US$104 million in the form of a results-based payment (RBP) was approved for Indonesia in August 2020 by the World Bank's Green Climate Fund (Global Forest Coalition 2020), in spite of civil society groups issuing an 'Open Letter to Members of the Green Climate Fund (GCF) Board' signed by more than 80 organisations demanding the GCF Board refrain from approving more REDD+ projects.

> The GCF is paying governments for supposed results arising from reduced deforestation in the past. These emission reductions are likely to exist on

Justice in the age of the Anthropocene 153

paper only: the governments of Indonesia and Colombia chose the period for which they claim reductions. They also set the reference levels against which actual deforestation is compared during the time they claim to have reduced deforestation. This opens the door for skilful manufacture of calculations that will result in an outcome that is favourable to the respective country. For example, by using inflated reference levels, a country can calculate emission reductions from avoided deforestation even if deforestation rates are rising [as they are in Indonesia].

(Open Letter to Members of the GCF Board 2020)

These injustices point in the direction of other dimensions and claims for justice, distinct in their own right, though interlinked. That is, calls for recognition of adat tenure have a triple guise addressed below. This can also include representation through meaningful participation, that goes beyond procedural rules about gender mainstreaming of women's inclusion in decision-making processes regarding REDD+ that conventionally mandates quotas.

Political representation as 'meaningful participation' would acknowledge the subjective positions of forest-dependent women are taken into account, their specific social roles are acknowledged, their specific needs are not only listened to but heard by project developers, and are implemented. Meaningful participation and gender mainstreaming, both participatory processes analysed in the Indonesian study, have been viewed instrumentally, thus giving the appearance that procedural justice is met. However, instrumental participatory processes are inadequately equipped to deal with power differentials or to addressing the issues related to the recognition of customary communities (Mathur et al. 2014, p. 47). Thus, these injustices are not only of economic maldistribution, although that is heavily implicated, but of misrecognition and of misrepresentation.

Indonesian study – major findings: recognition of adat tenure and its ongoing implementation

One of the major findings of the Indonesian study was the widespread and consistent call from participants across all three sites in the Indonesian study for recognition of adat tenure. The focus on tenure issues (see Chapter 6) had the effect of overwhelming findings specifically relating to women, addressed in the women-specific findings chapter, 'Where are the women?' The aspiration of adat tenure is not gendered but a widely held aspiration of customary communities for self-determination through secure tenure. As the interrelated nature of injustices makes clear, claims for justice on one dimension do not cancel out claims for seeking justice on another. Indeed 'justice ... will not be fully reached without addressing justice in each realm' (Schlosberg 2004, p. 523).

Securing adat tenure has been an ongoing issue widely shared by customary communities, academics, national and international social justice and

154   *Justice in the age of the Anthropocene*

environmental non-governmental organisations (NGOs). This culminated in a landmark verdict by Indonesia's Constitutional Court in May 2013, ruling that indigenous peoples' customary forests should not be classed as state forest areas, thus confirming the existence of indigenous-owned customary forests (Decision Number 35/2012).

Notwithstanding the significance of this historic verdict, the second most salient finding to emerge from the study was the issue of implementation of the Court's verdict which is still an ongoing issue of procedural justice. In other words, legal recognition may be rhetorically acknowledged but has so far been ignored in political practice. What Fraser's framework tells us in this situation, and the adat communities know themselves, is that for adat communities to have recognition take effect, where their customary forests are redistributed on the basis of the Constitutional Court's verdict, then this becomes an issue for the political order, the Indonesian State and its multi-level government structure at the national, provincial, district and sub-district scales to align decision-making rules, procedures and legislation to reflect the Court's verdict. Thus, implementation is a call for political representation, where Fraser, at least, 'has been very clear in her arguments that recognition is an element of justice, to be considered against distributional and participation issues ... [that is] a triavelent conception of justice' (Schlosberg 2004, p. 521). In this way, achieving the implementation of secure adat tenure, assumes a triple guise of justice claims-making.

If adat tenure were secured, via full implementation of the verdict, would customary communities be willing to proceed with carbon mitigation programmes such as REDD+, a global mitigation programme whose design to mitigate GHG emissions is global in scale, but experienced at the local level? That, of course, is a hypothetical question, given that implementation of the Constitutional Court verdict has not been realised. And yet, there are indications to suggest that if cultural ways of forest conservation and preservation were respected, acknowledged and validated this could be a possibility (D Central Kalimantan 2014). Indeed, a representative from the Asia Indigenous Peoples Pact in Preparations for the 2014 World Conference on Indigenous Peoples, an NGO that is neither for nor against REDD+, stated that on the ground in Cambodia, for example,

> indigenous leaders had asked 'if REDD is going to stop the concessions, the government from giving away our land to concessionaires, or to rubber plantations, then we'd rather go for REDD, than anything else, because with REDD, if REDD will assure that our forests will remain standing and that we will be able to do our livelihoods, then that is the way for us to go'.
>
> (Carling 2013)

Full jurisdictional recognition of adat tenure, which is what a call for implementation of the Constitutional Court's verdict would deliver to adat communities in Indonesia, would enable those affected by carbon projects such as REDD+, or any other intervention, national or international, to make those decisions for themselves (D Central Kalimantan 2014). However, absent this recognition, that is interwoven with representation as demonstrated here, how then do forest-dependent adat communities participate as equal peers in decisions that affect the environmental and social integrity of their livelihoods and emotional, physical and cultural well-being?

Unfortunately, appeals to Fraser's spatially expansive 'all-subjected principle' of affectedness to overcome the issue of misframed justice at a meta jurisdictional level, evident in the disconnect between a global mitigation programme such as REDD+ and the host communities impacted by its presence, that is the justice claimants, cannot remedy these injustices of misrecognition and misrepresentation. And the reason for this is because what continues to lock in and dominate 'current international negotiations, is not the question of duties to global others but rather the entitlements of sovereign states and private capital to exhaust precious resources reserves for private gain' (Skillington 2012, p. 1196). Justice can be 'theoretically' unbound as Fraser's framework has attempted. However, absent the Indonesian state and various levels of government aligning to implement the verdict of the Constitutional Court to uphold these interlinked conceptions of justice, then their corresponding injustices will remain. In the final analysis, this places adat communities' claims for justice squarely in the national frame for their redress and realisation by the political order of the Indonesian state.

Environmental reintegration: the 'fourth dimension' and the injustice of maladaptation

What does the fourth dimension of justice, environmental reintegration, proposed here, mean? And how is one to understand its corresponding injustice of maladaptation? The Indonesian study sought, among other aims, to construct local understandings of what climate justice (environmental justice) means and to use insights gained from participant responses to inform, in a preliminary way, the contours of a fourth dimension of justice, environmental reintegration. Similar to the notion of gender justice, climate justice too was overwhelmingly understood by participants from the perspective of injustice. In other words, both gender justice and climate justice are defined by reference to what 'starts not so much from a clear-sighted definition of what justice [is] but from widely shared intuitions of injustice' (Barnett 2010, p. 248). As one respondent aptly noted in relation to what climate justice means in REDD+, 'we have to make the layers of injustice visible first' (A2 Jakarta 2014).

In laying out the parameters of this fourth dimension, the arguments employed are heavily indebted to the research contained in Schlosberg's (2004) seminal paper 'Reconceiving environmental justice: global movements and political theories'. In this paper, he employs Fraser's (2008) three-dimensional framework to claim that articulations for environmental justice by global and

156 *Justice in the age of the Anthropocene*

Southern NGO movements are inextricably and thoroughly linked. So much so, that

> [i]t is not simply that the justice of environmental justice in political practice includes issues of equity, recognition and participation; the broader argument here is that the movement represents an integration of these various claims into a broad call for justice.
>
> (Schlosberg 2004, p. 527)

Schlosberg's (2004) assessment about the environmental justice movement as representing an integration of Fraserian three-dimensional justice into a broad call for justice was mirrored in findings from the in-country Indonesian study. All participants interviewed in the study worked for NGOs advocating for adat communities in Indonesia, addressing multiple issues and overlapping claims for justice ranging from: poverty alleviation, improved livelihoods (redistribution); addressing marginalisation, self-determination, gender rights, cultural respect, human rights (recognition); and land rights, management rights, sustainable land use planning, legal acknowledgement from the regional provinces, human security, as well as providing information about REDD+ that could be understood in their terms to facilitate meaningful participation in project interventions (representation). Additionally, these various claims for justice overlap; for example, the land rights issue is not simply a call for representation, but for recognition as forest-dependent communities with specific culturally based knowledge systems for managing their forests and interconnected hydrological systems (D Central Kalimantan 2014). Similarly, having secure customary forest tenure would address many of the claims for redistribution, such as improved livelihoods and welfare needs, that the current benefit-sharing arrangements under REDD+, in its first 10-year review, 'remains unclear' (Angelsen et al. 2018, cited in Global Forest Coalition 2020).

In an attempt to reassess the Fraserian project to include a fourth dimension of justice this book calls 'environmental reintegration', does not strictly adhere to the format that has shaped Fraser's families of justice. That is, under the three-dimensional framework, obstacles to justice emerge from the political, economic and social spheres of society that 'give rise to conceptually distinct species of injustice … [where] there can be distinctively political obstacles to parity [say in the political constitution], not reducible to maldistribution [the economic order] or misrecognition [the social order], although interwoven with them' (Fraser 2008, p.18).

Can a conceptually distinct species of justice, reintegration, arise from the 'environment', that is not reducible to representation, recognition or redistribution? To follow Fraser's trajectory for thematising reintegration as a fourth dimension of justice, would require one to say the following: obstacles to reintegrative justice arise distinctively from the environmental or ecological sphere, as opposed to the political constitution of society (representation), the class structure (redistribution) or the social order (recognition). And, to

Justice in the age of the Anthropocene 157

continue this Fraserian mode of theorising the 'fourth dimension' of justice: reintegration would then be the defining issue of the environment, a meta-issue of the frame, and its characteristic injustice would be maladaptation.

However, the environment (or nature), does not have personal agency; it cannot speak; it cannot participate or interact freely as a peer in society, and so it cannot achieve parity of participation on Fraserian terms conceptually. That is, as a distinct species of justice claims-making. Agency, as arguments arrived at conceptually and empirically throughout this book, demonstrates reintegration of the environment is mediated through the political, economic and social orders of societies, all of which are deeply embedded in the construction of *environment as anthropogenic climate change*. For example, reintegration is mediated through the political order via the polluter pays principle (PPP) implicitly acknowledged in the Climate Convention through the fundamental principle of 'common but differentiated responsibilities' (CBDR). CBDR is contained in Article 3.1 of the UNFCCC which begins not with a statement about the distribution of the Parties' responsibilities, but with an unambiguous statement about protection of the environment that implicates human agency. That is that, 'The Parties should protect the climate system for the benefit of present and future generations of humankind ...' (UNFCCC 1992, Article 3.1). What Article 3.1 then acknowledges is the role human agency plays through the political orders of the Parties in protecting the climate system, thus ensuring that the ultimate objective of the Convention is upheld.

> The ultimate objective of this Convention and any related legal instruments that the Conference of the Parties may adopt is to achieve, in accordance with the relevant provisions of the Convention, stabilization of greenhouse gas concentrations in the atmosphere at a level that would prevent dangerous anthropogenic interference with the climate system. Such a level should be achieved within a time-frame sufficient to allow ecosystems to adapt naturally to climate change, to ensure that food production is not threatened and to enable economic development to proceed in a sustainable manner.
>
> (UNFCCC 1992, Article 2)

To suggest, then, that the 'environment' is a distinct obstacle to achieving reintegration would require a return to environmental determinism. That is, 'environmental (climatic) determinism implies that a society is formed and determined by the physical environment, especially the climate' (van Liessum 2011). However, as Head (2010), cited in Chapter 3, claimed, 'recognition of anthropogenic climate change ... was in many ways the final nail in the coffin of environmental determinism' (Head 2010, p. 236). Further arguing, 'that if humans and their activities are embedded in the very structure of the atmosphere, we needed new ways of thinking about things' (Head 2010, p. 236). Thus the environment *as* anthropogenic climate change is inextricably linked,

158 *Justice in the age of the Anthropocene*

not only to the political orders of the Parties adhering to the PPP for its protection to achieve reintegration, but implicates redistribution through the CBDR and the Parties' capabilities to ensure its ultimate protection.

What the ultimate objective of the Climate Convention then invokes is that reintegration requires that the political, facilitated by the Conference of Parties, ensures that the economic and social orders are all aligned. That is, in sync with the ultimate protection of environment – climate stability. Fraser (2008) then correctly diagnosed the third dimension to the framework theory of justice. That is, that it is the sphere of the political order that lays out the ground rules for claiming justice that arise from economic, social *and* environmental injustices. Indeed, findings from the Indonesian study overwhelming endorse this point. Securing adat tenure is but one example.

It is indeed relevant at this point for the reader to ask whether or not 'reintegration' is a necessary and valid extension of Fraserian justice as a 'fourth dimension', given, as has been suggested here, that the 'political order', represented by the State Parties, is the primary conduit through which environmental injustices are addressed in achieving the ultimate goal of climate stability. Of course, that question assumes, which this chapter does not, to follow Fraser's (2008) thematising justice as previously and strictly posited. That is, to establish a distinct species of justice emerging from the ecological sphere which would then make such a question relevant. However, the meaning for reintegration as suggested here, emerges from the environment in the context of anthropogenic climate change, which implicates human agency (anthropogenic), and is thus interconnected to these other spheres for its redress in the avoidance of the injustice of maladaptation. *Reintegration may then be interpreted from the broader point of view of avoiding the injustice of maladaptation.* In this nascent attempt to provide the parameters of a fourth dimension, defining 'reintegration' has endorsed what Barnett (2010) suggested earlier: it is easier to gain a clearer conception of justice by starting with widely shared intuitions of injustice (Barnett 2010, p. 248).

Indeed, early author intuition, supported by critical analysis throughout this book, unambiguously suggests that current adaptation and mitigation solutions are leading to maladaptation. That is, maladaptation arises from climate change actions, policies or responses that relative to, or in the absence of, alternatives, 'increase emissions of greenhouse gases; disproportionately burden the most vulnerable; have high opportunity costs; reduce incentives to adapt; and, set paths that limit the choices available to future generations' (Barnett & O'Neill 2010, p. 211).

Through a wide-ranging structural critique of current adaptation policies (see Chapter 3) and mitigation policies (see Chapter 4) under the UNFCCC, what is increasingly evident is the need to adopt maladaptation as a concept through which to analyse these policies for their negative consequences, spatially and temporally (Barnett & O'Neill 2010; Juhola et al. 2016; Magnan et al. 2016).

In the absence of maladaptation as a principle through which to guide climate change solutions, Magnan et al. (2016) offer the 'anticipation of the risk

Justice in the age of the Anthropocene 159

of maladaptation to become a priority for decision makers and stakeholders at large, from international to local levels' (Magnan et al. 2016, p. 646). However, the authors also noted their concern with a 'statement made by the IPCC's Fifth Assessment Report: 'Five dimensions of maladaptation were identified by Barnett & O'Neill [in 2010 that] are useful pointers to the potential for maladaptation but their application depends on subjective assessments' (Noble et al. 2014, p. 868, cited in Magnan et al. 2016). The framing of 'maladaptation' as subjective is an unsurprising statement from the world's scientific authority on climate change, but its framing of climate change as gender neutral offers a cautionary tale that omission via scientific positivism has cumulative flow-on effects that do not reflect on-the-ground realities. At a minimum, the IPCC has the capacity to refer the issue of maladaptation as a precautionary measure, to avoid harm, to the UNFCCC COP for further scoping, and the empirical work on accumulating experiences of climate change solutions, adaptation policies in particular, leading to maladaptation empirically demonstrate this (Juhola et al. 2016; Magnan et al. 2016; Vigil 2015, 2018; Work et al. 2019).

This book suggests a proxy to avoiding maladaptation in adaptation policies is found in the precautionary principle (see Chapter 3). However, the principle under the Climate Convention has been circumscribed around mitigation, that is 'the Parties should take precautionary measures to anticipate, prevent, or minimise the causes of climate change and mitigate these effects' (Article 3.3, UNFCCC 1992). Following the analysis of structural flaws inherent in international crediting mechanisms producing negative outcomes, notably environmental integrity in the CDM (Cames et al. 2016), as well as in REDD+ (Global Forest Coalition 2020; Open Letter to Members of the GCF Board 2020), what we are left with is a maladaptive structure meant to mitigate; how, then, does adaptation policy adapt to maladaptation resulting from 'ongoing efforts to "fix" and resuscitate failing market-based mechanisms' (Delabre et al. 2020, pp. 4–5).

Precaution, as a more general principle, should be extended to adaptation policies to avoid, for example, disproportionately burdening the most vulnerable to the impacts of climate change. Evidence from desk-based conceptual research in Chapter 3, suggests that the dominant approach to adaptation policy, typified as 'adaptation plus development' is an approach that should be discarded (Brooks et al. 2009). That is because this approach calls for 'tacking on' adaptation to mainstream development agendas, evident in terms such as 'climate proofing' development investments and 'mainstreaming' adaptation into development, thus leaving 'development' intact (Ayers & Dodman 2010). However, as Huq et al. (2006) correctly claimed, the underlying cause of climate change is unsustainable development. Thus, what 'adaptation plus development' reinforces is that economic growth, modernity and progress, articles of neoliberal economism, driving current development approaches remain uninterrupted. Indeed, Work et al. (2019) found in their research in Cambodia that climate change mitigation and adaptation policies themselves are a driver of potential maladaptation, 'entwined as they are with an

160 *Justice in the age of the Anthropocene*

unchanged vision of development and the policy frameworks that support it, [and that] give rise to projects that are largely only discursively different from the development projects of the past' (Work et al. 2019, p. 58).

Similarly, another approach to adaptation policy, 'sustainable adaptation', that is, where adaptation policies must meet a dual threshold of social justice and environmental integrity are, according to Brown (2011), leading to maladaptation. The reason for Brown's (2011) claim that this interpretation of adaptation is leading to maladaptation, is due to sustainable adaptation as being nothing more than an oxymoron because it shares similar features to sustainable development, 'a deliberately vague and slippery term' that fosters, among other things, the notion of sustainable growth. In fact, as ecological economists Daly and Townsend (1993) noted close to three decades ago, sustainable growth falls in the ranks of impossibility statements. That is,

> it is impossible for the world economy to grow its way out of poverty and environmental degradation. In other words, sustainable growth is impossible. In its physical dimensions the economy is an open subsystem of the earth ecosystem, which is finite, non-growing, and materially closed. As the economic subsystem grows it incorporates an ever greater proportion of the total ecosystem into itself and must reach a limit at 100 per cent, if not before. Therefore, its growth is not sustainable. The term 'sustainable growth' when applied to the economy is a bad oxymoron – self-contradictory as prose, and unevocative as poetry.
>
> (Daly & Townsend 1993, p. 267)

Chapter 4 used Barnett and O'Neill's (2010) maladaptation typology to screen the Clean Development Mechanism for its potential to lead to maladaptation. There, in a lengthy and in-depth critical analysis of the CDM, indications are it is leading to maladaptation along all five pathways. That is, firstly through increased greenhouse gas emissions (Cames et al. 2016; Evans & Gabbatiss 2019b; Liverman 2009). Secondly, the CDM has created a disproportionate burden on developing countries in favouring the more developed of the developing countries, 'leaving the bulk of the South on the side-lines of the global carbon market' (Banuri & Gupta 2000, cited in Najam et al. 2003, p. 225; Evans & Gabbatiss 2019b). Thirdly, the high opportunity cost of adopting the CDM, relative to alternatives such as a regulatory carbon tax among others, is a lost opportunity. Fourthly, an action that reduces incentives to adapt as a pathway to maladaptation was similarly bound to the previous opportunity cost of alternative command and control mechanisms, for example, environmental regulations, carbon taxes and the like being adopted, rather than an international crediting mechanism such as the CDM. Here, the profit motive is at the heart of the neoliberal argument for emissions trading and offset mechanisms, 'accumulation by decarbonisation' (Bumpus & Liverman 2008), thus reducing incentives to adapt to these alternative mechanisms, that is, business as usual. And finally, neoliberal environmental solutions to climate

Justice in the age of the Anthropocene 161

change have 'set paths that limit the choices available to future generations' (Barnett & O'Neill 2010, p. 211).

In Chapter 5, previously discussed in light of Fraser's theory of justice, Mellor's (2012) model of the Economic Dualism has provided other insights to adaptation and mitigation policies: for example, the ongoing failure to reintegrate the other half of the Economic Dualism, where mostly women's work, indigenous communities and nature reside, is a form of maladaptation. That is because technical carbon-centric policies focusing on mitigating greenhouse gases have come at the expense of women's subjective needs being adequately addressed as already demonstrated; REDD+ has side-lined alternative mitigation methodologies advanced by adat communities that rely on indigenous knowledge in protecting their forested lands and environment, such as community-monitoring, where 'nature' is defined narrowly in the context of mitigating greenhouse gas emissions. The Indigenous Organisations of the Amazon River Basin (COICA), in recognising the targeting of indigenous lands for climate change mitigation strategies, approached researchers at the Woods Hole Research Centre to conduct analysis on carbon storage within indigenous territories and protected natural areas (Walker et al. 2015, cited in Osborne 2018, p. 64). Findings of the study revealed that indigenous peoples of the Amazon 'played an important role in forest stewardship, and that their territories are associated with low levels of deforestation and are responsible for storing nearly one third of the region's aboveground carbon' (Walker et al. 2015, cited in Osborne 2018, p. 64).

In summary, current interpretations of adaptation and mitigation solutions critiqued in this book are framed within the logic of market environmentalism and its associated privileging of market-based solutions, 'a quintessentially neoliberal strategy for addressing climate change' (Osborne 2018, p. 62). This privileging has left local-level actors, such as the adat communities the Indonesian study investigated, with insufficient influence at the national and international levels of climate change negotiations to advance their claims and protect their interests (Mathur et al. 2014), and thus the environments in which they live where carbon-based projects are located.

(2) Are climate change solutions leading to maladaptation – representing a different trigger for environmentally induced migration and displacement?

The analyses of both desk-based conceptual research and empirical findings from the Indonesian study, demonstrate that adaptation and mitigation policies under the UNFCCC are leading to maladaptation, thus presenting as a different trigger for environmentally induced migration and displacement, in addition to climate change itself. Barnett and O'Neill's (2010) typology of maladaptation has been used as a conceptual intervention, as others have invoked the 'term maladaptation to intervene in the persistent optimism of [mitigation] policy driven solutions' (Mosse 2005, cited in Work et al. 2019, p. 51). On all five pathways through which maladaptation arises under this guide, the

162 *Justice in the age of the Anthropocene*

CDM appears to distinguish itself as a mechanism for maladaptation and not mitigation.

The question here is, has the CDM led to migration and/or displacement of peoples where CDM projects are located, that is in the global South? To anticipate the answer, yes; and the analysis in support of this argument relates to the concept of 'additionality', which no CDM project would be allowed to progress without. Article 12.5 of the Kyoto Protocol outlines what constitutes additionality over two key clauses. That is, 'emission reductions resulting from each [CDM] project activity shall be:

a Real, measureable and long-term benefits related to the mitigation of climate change; and
b Reductions in emissions that are additional to any that would occur in the absence of the certified project activity.'

(UNFCCC Kyoto Protocol 1998)

In other words, additionality distinguishes reductions in emissions generated by the CDM project from the business-as-usual baseline emissions without the project (Michaelowa 2005, cited in Bumpus & Liverman 2008). It thus relies on describing counterfactual (non-existing) future scenarios by CDM project developers which can lead to two types of additionality not being met: environmental and financial additionality. For example, in the case of no environmental additionality being met, this occurs when baselines are exaggerated and claims of reductions in emissions have not actually occurred (Cames et al. 2016; Evans & Gabbatiss 2019a, 2019b). With respect to no financial additionality being met by the proposed project, this occurs when carbon credits, that is, Certified Emission Reduction (CER) credits are generated that would have happened anyway (Bumpus & Liverman 2008). These projects should therefore not qualify under the CDM because they do not create additional emission reductions. In fact, they exacerbate the problem of climate change by allowing companies in the developed world, as the purchasers of CDM CER credits, to exceed their emission limits without genuinely offsetting elsewhere.

Chapter 4 presented evidence, early in the first commitment period of the Kyoto Protocol (2008–2012), in support of spurious emission reductions under the CDM, described at this time as 'alarming' (Pearce 2008). In 2007, an expert adviser of the CDM executive board said one-third of Indian CDM projects lacked additionality or were simply business as usual (Smith 2008; Lohmann 2009). David Victor from Stanford University claims emission reductions have not resulted from over two-thirds of CDM projects in developing countries that have been issued with credits (Wysham 2008). The pattern continued through the second commitment period as well (2013–2020), where Cames et al.'s (2016) empirical review of whether the CDM was meeting additionality in the second commitment period found 'the CDM has still fundamental flaws in terms of environmental integrity. It is likely that the large majority of

Justice in the age of the Anthropocene 163

the projects registered and CER issued under the CDM are not providing real, measureable and additional emission reductions' (Cames et al. 2016, p. 11).

In another early study by International Rivers, an international NGO that campaigns against hydroelectric dams, the vast majority of the many hundreds of hydroelectric dams under construction in China issued with CERs were either under construction or completed before the application of carbon credits were made (Pearce 2008; Smith 2008, 2009). In other words, the suggestion here is the projects were going to be developed anyway, regardless of finance (Smith 2008). That is, regardless of carbon funding, the carbon projects would have been implemented in any case (Osborne 2018). Cames et al. (2016) recommended 'excluding large scale hydropower projects from being eligible under the CDM, due to overall questionable additionality' (Cames et al. 2016, p. 112).

If, on the analysis presented here, bogus emissions have been made and CERs issued from CDM projects – for example, the hundreds of hydroelectric dams under construction in China – then displacement resulting from this kind of development would usually fall under planned development projects, whose major driver would be categorised as 'development forced displacement and resettlement (DFDR)' (Bronen 2009, p. 15).

However, displacement of peoples under the scenario presented here should more accurately be categorised as displacement due to maladaptation inherent in the CDM. Notwithstanding the 'overall questionable additionality' of large-scale hydropower projects under the CDM (Cames et al. 2016), which would require suspending disbelief in their environmental integrity, if emission reductions have been claimed and developed countries have purchased the CERs generated to meet their Kyoto obligations (or potentially, 'voluntary' Nationally Determined Contributions under the Paris Agreement's new mitigation mechanism), it is technically correct to designate maladaptation as the trigger for migration and/or displacement. If this proposition holds, further research would be needed to gauge the extent of maladapted migration under the CDM – and its post-Kyoto successor established under Article 6 of the Paris Agreement.

In Chapter 6's general findings of the Indonesian study, migration and/or displacement of peoples as a result of the REDD+ pilot project, KFCP in Central Kalimantan and REDD+ readiness activities in Central Sulawesi were more ambiguous for reasons presented below. Based on findings in Central Kalimantan, one resident respondent was not aware of any displacement resulting from the KFCP project (D Central Kalimantan 2015). However, another respondent claimed that a REDD+ project in East Kalimantan left many in the community questioning whether they could still gain access to the forest.

> Usually they can freely go into the forest for instance for wood … people are still afraid for asking themselves whether 'it is still safe to go in there, is it OK?' I mean there are projects there, there are guards there, there are

164   *Justice in the age of the Anthropocene*

people there … maybe not like that military type but there are people there from the project, oh yeah, there are people like that, intimidated to go and do the usual stuff that they do inside the forest.

(N Jakarta 2014)

SND: They just want the forest for the REDD conservation [project area undisclosed] and a lot of the access is closed … the conservation area they always have regulation, every tree, every leaf that they take from the forest for their family … sometimes the police will [inaudible] for crossing the forest … AG: … And if they do? S: Then they really go to gaol.

(SND Jakarta 2014)

At the time of commencing the in-country study in April 2014, 'numerous REDD+ projects noted on the Ministry of Forestry website, [were] underway in various provinces in Indonesia' (AP2 Jakarta 2014). As of May 2018, globally there were 350 REDD+ projects in 53 countries, Indonesia with 21 projects underway, placing it fourth out of 10 key countries that host the most REDD+ projects, surpassed by Brazil with 48, Colombia 33, and Peru 25 projects (Angelsen et al. 2018, cited in Global Forest Coalition 2020, p. 4).

Given the multiple REDD+ projects underway in Indonesia, more specifically, in light of the previous participant responses, one has to question the extent to which adat communities' human security, freedom from want, freedom from fear are routinely violated. Although no evidence of physical displacement/migration had resulted due to REDD+ in the two provinces under study, as the previous responses indicate, there is clear evidence of displacement in terms of accessing forests for livelihood provisioning. Vigil's (2018) research on green-grabbing displacement supports this finding, calling it 'in situ displacement [which] occurs when REDD+ prevents communities from accessing forest areas and diminishes the area of cultivate land, with impacts on the livelihoods and food security of forest dependent communities unable to find alternative livelihoods' (Vigil 2018, p. 377).

The REDD+ programme in Central Sulawesi was at the 'readiness' phase of the programme at the time of the in-country study (June 2014). That is, unlike Central Kalimantan, no on-the-ground demonstration activities had been implemented. However, movement into this phase of REDD+ was imminent with the near-completion of REDD+ readiness activities. Thus, it was not possible to determine whether migration or displacement would occur – due to the temporal dimension of maladaptation. However, one scenario of potential 'in situ displacement' of adat communities was raised by several of the respondents, which centred on a potential extension of Lore Lindu National Park's boundaries to accommodate the REDD+ demonstration activities project area. One village participating in the FPIC readiness activity is located in the sub-district of Gumbasa which shares a border with the national park (SB Central Sulawesi 2014). However, at the time of the study no decision had been made as to the possible expansion of Lore Lindu to accommodate REDD+.

Justice in the age of the Anthropocene 165

Finally, as evidence from the findings suggested, REDD+ is considered as just one more project in a long history of many interventions into adat communities' forests. The main concern respondents expressed across all three research sites is the issue of securing adat tenure to stop the concessionaires, oil palm plantations, logging and mining companies who represent the real drivers of deforestation in Indonesia (see Delabre et al. 2020; Maxton-Lee 2018, 2020; Open Letter to Members of the GCF Board 2020). They 'want recognition for their titles and their rights to everything in their territory. Basically they want their rights to be respected meaning that they [the State] don't violate them. I mean legally acknowledged by the State' (AP1 Jakarta 2014). Notwithstanding the perverse possibility that concession holders may be able to gain access to REDD+ projects.

That is, 'there is the Forestry Minister Regulation No.20/2012 about the commodification of forest carbon. That regulation was very technical, [but] it gives title to implement forest carbon business to existing concession holders' (AP2 Jakarta 2014). However, adat communities do not have carbon rights because the Constitutional Court ruling has yet to be implemented which, of course, would give them title. The Ministry of Forestry has asserted through regulation No.20/2012 that

> logging concession holders, even palm oil concession holders can apply for REDD+ projects if they, for example, change the [inaudible] methodology, Oh! I'm doing this sustainable forestry management, give me REDD+ money. Oh! I'm developing a forest plantation, 2 million hectares, if it absorbs carbon more than the natural forest, give me REDD money.
>
> (AP2 Jakarta 2014)

When questioned whether this has happened as yet, the respondent answered affirmatively, claiming s/he had attended many seminars and workshops with, for example, the Association of Palm Oil Plantations in Indonesia and Forest Plantation. What they are trying to do is, 'they are trying to appropriate REDD+' (AP2 Jakarta 2014). In essence, these bodies have been pushing to change what the definition of a forest is. For example, Forest Plantation claimed that natural forests do not absorb as much carbon over a period of seven years as a new forest plantation would (AP2 Jakarta 2014). 'Although plantations can store carbon relatively rapidly they provide none of the benefits that real forests provide' (Global Forest Coalition 2020). The Global Forest Coalition is an international coalition of indigenous peoples' organisations and NGOs defending social justice and the rights of forest peoples in forest policies. A recently published 15-year review of REDD+ cites a recent study 'that showed natural forests are 40 times better at storing [carbon] than tree plantations' (Lewis et al. 2019, cited in Global Forest Coalition 2020, p. 2).

Similarly, the Palm Oil Association has sought to redefine palm oil as a forest and not as a crop. However, the attempt at redefining the latter

166 *Justice in the age of the Anthropocene*

succeeded for a week before the Ministerial Regulation was revoked due to massive protests. Palm oil is not forest ... and it requires actually destroying a whole ancient ecosystem to make room for that and it requires great burning ...

(AP2 Jakarta 2014)

Had the Ministerial Regulation not been revoked, the potential to create displacement of forest-dependent peoples in Indonesia would have increased. However, the Indonesian Ministry of Forestry excludes forest plantations from the deforestation definition. In other words, if a natural forest is destroyed, the Ministry does not count this as deforestation if it is for forest plantations. 'So that is one battle we have yet to win ... and the target is 20 million [hectares] ... [we can] expect grabbing due to forest plantation and [the] forest plantation is trying to be sold for REDD+ to appropriate REDD+ money' (AP2 Jakarta 2014). And the green-grabbing induced displacement of forest-dependent peoples that Vigil (2018) calls to be included in the work of migration scholars. These responses clearly demonstrate that through a regulatory sleight of hand, the State with powerful resource companies' interests involved, can facilitate the prospect of displacement due to maladaptation inherent in the REDD+ scheme, which may yet come to pass (see also, Work et al. 2019).

(3) What is the basis of just adaptation and mitigation polices that can be adopted by the UNFCCC once these have been subsumed within a global climate carbon market framework – of its own making?

It is clear that the key question addressed through conceptual research and empirical findings in this book could equally have been inverted to ask: What mitigation and adaptation policies can be adopted by the Conference of Parties to the UNFCCC to avoid the injustice of maladaptation? As the analysis has shown, it is through an articulation of injustices that a range of just possibilities or alternatives to the dominant neoliberal market environmentalism for dealing with climate change can emerge. These alternatives counter, in modest and more just ways, the current carbon-centric and market-driven logic that prevails under the global carbon market framework the Kyoto Protocol heralded in.

A structural analysis of market-based mechanisms such as the CDM and REDD+ have locked in a path of dependency, while foreclosing on alternative pathways to more just adaptation and mitigation policies. Such alternative policies can be directed under the UNFCCC and adopted by the Conference of Parties. However, this chapter has shown these alternative solutions can *only* be enforced by State Parties. That is because Fraser's 'all-subjected' principle under conditions of globalisation cannot be implemented due to the obstacles present in the political orders of State Parties, where claims for justice by those affected by climate change solutions are mediated through the state's legislature. Further, it is difficult to imagine that nation-states, powerful states in

Justice in the age of the Anthropocene 167

particular, 'would willingly compromise territorial sovereignty in order to pursue a collective global climate change polity based on equity and justice' (Kythreotis 2012, p. 460). As desirable as this may be in a world where 'climate change does not respect national borders' (Ki Moon 2015), states, and their sovereignty, are not going anywhere fast.

Consequently, the alternative solutions presented here may usefully be conceptualised on a continuum of 'what is possible' to 'what ought be', the latter discussed in the 'Afterword' at the end of this chapter entitled: 'A questioning moment: a conjunctural crisis manifest in Extinction Rebellion'.

We already know from Huq et al. (2006) that unsustainable development is the underlying cause of climate change. And unsustainable development is rooted in an unsustainable neoliberal economy that drives development trajectories towards ever-expanding unsustainable growth. It is also the dominant neoliberal economic paradigm through which the solutions to climate change, mitigation and adaptation policies, have been framed as an accepted methodology for dealing with climate change (Bailey & Wilson 2009).

> Carbon commodification [CDM, emissions trading, REDD+] can, thus be seen as a partial assimilation of a new idea (the need to curb human interference with the climate system) into the prevailing paradigm, with framing of its solutions in the context of pre-existing discourses (the promise of economic efficiency and environmental effectiveness) (Bakker 2005; Heynen 2007b), rather than the adoption of less conventional strategies that focus, for example, on economic relocalisation or direct behavioural constraints to reduce fossil-fuel consumption and greenhouse gas emissions.
>
> (Hopkins 2008, cited in Bailey & Wilson 2009, p. 2335)

When Fraser (2008) first started developing her theory of justice, as a 'bivalent' conception incorporating recognition and redistribution (Schlosberg 2004, p. 519), the idea was to overcome the 'vulgar economism' that tended to dominate previous left politics. However, by the 1990s, with the rise of free-market neoliberalism, it became clear that in 'analysing the shift in the grammar of claims-making from "redistribution to recognition" I argued we should resist the displacement of redistribution by recognition by developing an integrated conception of justice that could encompass defensible claims of both types' (Fraser 2004, cited in Dahl et al. 2004, p. 376). The reason for returning to Fraser's earlier thinking about analysing the claims of justice, as not being on one dimension or the other, is because the market-based solutions to climate change analysed here are firmly entrenched in the 'vulgar economism' Fraser sought to disentangle.

However, as this chapter has argued, it is in the political sphere where claims for redistribution, recognition and representation are mediated and addressed. One could argue the emphasis on the political sphere represents a form of 'vulgar sovereignty'. However, alternative just solutions to climate

168 *Justice in the age of the Anthropocene*

change can only emerge within a framework of 'what is possible', and which positions the political order as the prime obstacle or conduit through which to realign with the other spheres of society to bring about reintegrative justice. That is, where reintegration is interpreted from the point of view of avoiding the broader injustice of maladaptation.

What is possible may deliver justice

The alternative mitigation and adaptation solutions to climate change proposed here are not impossible, though politically contingent on what has already been proposed by the Conference of Parties, for example, in REDD+, through agreed co-benefits and safeguards. It is not to suggest that these alternative pathways, such as mitigating climate change through, say, adat community monitoring, or even the notion of 'precautious adaptation', can deliver, on their own, the solution to climate change. In fact, 'climate change presents a challenge that will never be "solved" … we can [only] do better or worse in our managing of it' (Prins et al. 2010, p. 36).

Firstly, the inclusion of alternative customary peoples' methodologies, cultural practices, knowledge and systems for managing forests and natures would represent more just climate change solutions through the expansion of adat participation (Burrows 1997, cited in Schlosberg 2004). These strategies and practices have the potential to avoid the further injustice of misrepresentation, misrecognition and maldistribution. Hence, the need is for adat tenure to be immediately implemented. That requires the Indonesian state to proceed with implementation of the Constitutional Court ruling.

Currently, 'AMAN (the Indonesian Indigenous Peoples Alliance of the Archipelago) has a petition asking the House of Representatives to adopt a Bill on Recognition and Protection of the Rights of Indigenous Peoples' (Lang 2016). With respect to REDD+ this would ensure that REDD+ is genuinely 'rights-based' and participatory (Beymer-Farris & Bassett 2012).

Secondly, and more problematically, the drivers of deforestation are not linked to carbon under REDD+ (Okereke & Dooley 2010). However, as already discussed, Forest Plantation and the Palm Oil Association 'are trying to appropriate REDD+' (AP2 Jakarta 2014). It is possible for the Indonesian government through legislation, as it did with the Palm Oil Association, to ensure that the drivers of deforestation do not become the recipients of REDD+ money. Further, land tenure security and title for adat communities becomes doubly crucial to ensure this does not eventuate.

Thirdly, it is evident that REDD+ is a delaying tactic to avoid the hard choices that the Conference of Parties must make. When Norway established the $1 billion REDD scheme for Indonesia to reduce deforestation in 2010, six years later, of that money, only $60 million dollars had been handed over (Lang 2016). In a recent trip to Indonesia in March 2016, the Norwegian Climate and Environment Minister acknowledged that Indonesia has failed to reduce deforestation. 'We would obviously have hoped things would have

Justice in the age of the Anthropocene 169

progressed more quickly. We haven't seen actual progress in reducing deforestation' (Helgeson 2016, cited in Lang 2016). Indeed, Norway is a major oil producer and the money for REDD+ is generated by the very carbon emissions it seeks to reduce in Indonesia through REDD+. That is, 'if Norway buys REDD credits from Indonesia to allow continued drilling of oil ... it will at best move the emissions source from Indonesia to Norway' (Lang 2016).

Fourthly, as is apparent from the above, counterfactual scenarios, emission baselines before and after the project, central to both the CDM and REDD+ 'makes one shout out: Impossible! Fraud! Bribery! Corruption! Wasteful diversion of resources into pointless attempts at verification!' (Buiter 2007, cited in Lohmann 2009, p. 182). In this way, the concept of 'additionality' so central to CDM and REDD+ mitigation mechanisms falls into the realm of Daly and Townsend's (1993) impossibility theorems.

Conclusion

Finally, the issue of deforestation as a major driver of climate change and loss of livelihoods for those where it takes place is a deeply complex issue, sustained by myths and lock-ins (Delabre et al. 2020), ultimately connected to a global development trajectory, where the alternatives presented have been circumscribed around 'what is possible', rendering justice politically contingent. 'What ought be?' manifest in Fraser's (2008) theory of justice cannot be delivered in practice while the current and conventional faith in neoliberal environmental solutions prevails, where '[a] number of commentators have [even] questioned the possibility of reconciling the pursuit of justice with neoliberal environmental governance' (Lohmann 2008, Okereke 2008, cited in Mathur et al. 2014, p. 44).

However, hope is a necessary illusion in the pursuit of justice; un-remedied injustices for women and customary communities in the Indonesian study are testament to this. That does not imply they are without agency. On the contrary, it was through the coalition of indigenous peoples in Indonesia, represented by AMAN, that the issue of securing adat tenure was brought to the Constitutional Court and won. Full implementation is the next phase. What is known is that customary communities, where climate change projects are located, will at least have the self-determining right to decide, on their own terms, based on their own titled tenure, whether they pursue these projects, rather than have them imposed.

Afterword: A questioning moment: a conjunctural crisis manifest in Extinction Rebellion?

In the context of the ongoing effects of the 2008 global financial and banking crisis, Doreen Massey (2011) suggested that the contemporary global economy may be in crisis; however, the crisis appears insufficient for there to be a shift in the balance of social forces. For that to occur there must be a 'conjunctural crisis' – that is, when the political, ecological, cultural and economic forces are

170 *Justice in the age of the Anthropocene*

simultaneously aligned in a 'questioning moment'. Hall (2011) says they occur 'when a number of contradictions at work in different practices and sites come together – or "con-join" – in the same moment and political space' (Hall 2011, p. 9). Over a decade has passed and a new global health crisis has emerged with the onset of the COVID-19 pandemic.

John Gray (2020) says 'the virus has exposed fatal weaknesses in the economic system that was patched up after the 2008 financial crisis', declaring 'the hyper globalisation of the last few decades is not coming back … Liberal capitalism is bust' (Gray 2020). David Harvey (2020) states the existing model of capital accumulation was already 'in a lot of trouble before the virus', increasingly resting on fictitious capital (for example, 'silent offsets' under 'accumulation by decarbonisation'), debt creation and a vast expanse in the money supply. 'If I wanted to be anthropomorphic and metaphorical about this, I would conclude that Covid-19 is nature's revenge for over forty years of nature's gross and abusive mistreatment at the hands of violent and unregulated neoliberal extractivism' (Harvey 2020). For Nancy Fraser (2020), 'I would say coronavirus in the age of neoliberalism is a textbook lesson of the absolute imperative for a socialist feminist reorganization of society', claiming that what the virus has shown us is, 'it's not just that the production system depends on the care work, the care work depends on the production system'.

How would Massey (2011) have read the present situation, 'as a crisis, another unresolved rupture of that conjuncture that we can define as "the long march of the Neoliberal Revolution"' (Hall 2011, p. 9), or has there been a sufficient shift in the balance of social forces to bring about a new conjuncture?

'Conjunctural crises are never solely economic' (Hall 2011, p. 9), and Gray's, Harvey's and Fraser's calls on the present situation confirm this, by speaking to a wider range of social forces in simultaneous rupture. Gray (2020) marks the crisis as 'a turning point in history', declaring the political and economic 'era of peak globalization is over'. Harvey (2020) and Fraser (2020) highlight how the virus has illuminated what the status quo of (now broken) neoliberalism has rendered invisible: nature and social reproduction. And while a full conjunctural analysis is beyond this Afterword, its conceptual introduction is an opportunity to posit what this 'new conjuncture' may look like. The 'Pivot to Extinction' diagram in Figure 8.1 depicts how our choices are now limited by the existential threat of the climate crisis.

The 'Status Quo' (Iceberg/Layer Cake with Icing/Economic Dualism) must be turned on its side (in the direction of the upward arrows) to a position of Equilibrium. Hierarchy is then eliminated. The arc between the Status Quo and Equilibrium is representative of the alternatives that can facilitate the transition to a new state of equilibrium, in so doing bringing the political, economic, cultural and environmental to a position of alignment. For example, degrowth, steady-state economy, zero growth, diverse economies, reproductivity: affective labour, synergistic economies, re-commoning, decommodification of land (McAfee 2014).

If this ontological shift does not happen, another existential one will – a full pivot to X (Extinction). In this position, the diagram represents the egg-timer

Justice in the age of the Anthropocene 171

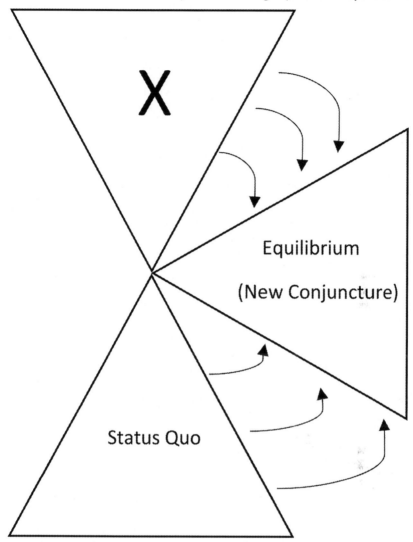

Figure 8.1 Pivot to Extinction (author's own diagram)

logo of Extinction Rebellion (XR) – a politically non-partisan international movement launched on 31 October 2018 in 'response to the IPCC Report we only had 12 years to stop catastrophic climate change and our understanding that we have entered the 6th mass extinction event' (rebellion.earth/the-truth/faqs). A more optimistic rendition than a full pivot to X, is the pressure XR are prepared to exert through an open 'strategy … of non-violent, disruptive civil disobedience – a rebellion' (extinctionrebellion.uk/the-truth/demands/). The downward arrows from X to Equilibrium are representative of a broader global environmental activism for climate justice.

172 *Justice in the age of the Anthropocene*

Notes

1 'In 2000 Paul Crutzen, an eminent atmospheric chemist, realised he no longer believed he was living in the Holocene [the existing paleontological era]. He was living in some other age, one shaped by people. From their trawlers scraping the floors of the seas to their dams impounding sediment by the gigatonne, from their stripping of forests to their irrigation of farms, from their mile-deep mines to their melting of glaciers, humans are bringing about an age of planetary change. With a colleague, Eugene Stoemer, Dr Crutzen suggested this age be called the Anthropocene – "the recent age of man"' (The Economist 2011, p. 81).

2 'The phrase "Keynesian-Westphalian frame" is meant to signal the national-territorial underpinnings of justice disputes in the heyday of the postwar democratic welfare state, roughly 1945 to the 1970s. The term "Westphalian" refers to the Treaty of 1648, which established some key features of the modern international state system. However, I am concerned neither with the actual achievements of the Treaty nor with the centuries-long process by which the system it inaugurated evolved. Rather, I invoke "Westphalia" as a political imaginary that mapped the world as a system of mutually recognising sovereign territorial states. My claim is that this imaginary informed the postwar framing of debates about justice in the First World, even as the beginnings of a post-Westphalian human-rights regime emerged' (Fraser 2008, p. 160, footnote 1).

References

Adger, W, Paavola, J & Huq, S & Mace, M (eds) 2006, *Toward justice in adaptation to climate change*, MIT Press, Cambridge, Massachusetts.

Ayers, J & Dodman, D 2010, 'Climate change adaptation and development I: the state of the debate', *Progress in Development Studies*, vol. 10, no. 2, pp. 161–168.

Baier, A 1991, *A progress of sentiments: reflections on Hume's Treatise*, Harvard University Press, Cambridge, Massachusetts.

Bailey, I & Wilson, G 2009, 'Theorising transitional pathways in response to climate change: technocentrism, ecocentrism, and the carbon economy', *Environment and Planning A*, vol. 41, no. 10, pp. 2324–2341.

Barnett, C 2010, 'Geography and ethics: justice inbound', *Progress in Human Geography*, vol. 35, no. 2, pp. 246–255.

Barnett, J & O'Neill, S 2010, 'Maladaptation', *Global Environmental Change*, vol. 20, pp. 211–213.

Bennholdt-Thomsen, V & Mies, M 1999, *The subsistence perspective: beyond the globalised economy*, Zed Books, London UK & New York, USA.

Bessis, S 2003, 'International organisations and gender: new paradigms and old habits', *Signs: Journal of Women in Culture and Society*, vol. 29, no. 2, pp. 633–647.

Beymer-Farris, B & Bassett, T 2012, 'The REDD menace: resurgent protectionism in Tanzania's mangrove forests', *Global Environmental Change*, vol. 22, no. 2, pp. 332–341.

Bronen, R, Chandrasekhar, D, Amor Conde, D, Kavanova, K, Morinière, LC, Schmidt-Verkerkand, K & Witter, R 2009, 'Stay in place or migrate: a research perspective on understanding adaptation to a changing environment', in A Oliver-Smith & X Shen (eds), *Linking environmental change, migration and social vulnerability*, UNU Institute for Environment and Human Security (UNU-EHS), e-book, accessed 21 May 2017, www.munichre-foundation.org/dms/MRS/Documents/Source2009_OliverSmith_ShenEnvironmentalChange_Migration.pdf.

Brooks, N, Grist, N & Brown, K 2009, 'Development futures in the context of climate change: challenging the present and learning from the past', *Development Policy Review*, vol. 27, no. 6, pp. 741–765.

Brown, K 2011, 'Sustainable adaptation: an oxymoron?', *Climate and Development*, vol. 3, no. 1, pp. 21–31.

Bumpus, A & Liverman, D 2008, 'Accumulation by decarbonisation and the governance of carbon offsets', *Economic Geography*, vol. 84, no. 2, pp. 127–155.

Cameron, J & Gibson-Graham, J 2003, 'Feminising the economy: metaphors, strategies, politics', *Gender, Place and Culture: A Journal of Feminist Geography*, vol. 10, no. 2, pp. 145–157.

Cames, M, Harthan, RO, Füssler, J, Lazarus, M, Lee, CM, Erickson, P & Spalding-Fecher, R 2016, *How additional is the Clean Development Mechanism?: analysis of the application of current tools and proposed alternatives*, Study prepared for DG CLIMA, Öko Institut, accessed 7 August 2020, https://ec.europa.eu/clima/sites/clima/files/ets/docs/clean_dev_mechanism_en.pdf.

Carling, J 2013, *Preparations for the 2014 World Conference on Indigenous Peoples – press conference, 28 May 2013: as part of the ongoing session of the Permanent Forum on Indigenous Issues*, online video, accessed 10 June 2017, http://webtv.un.org/watch/p reparations-for-the-2014-world-conference-on-indigenous-peoples-press-conference/2 416140319001/.

Carvajal-Escobar, Y, Quintero-Angel, M & Garcia-Vargas, M 2008, 'Women's role in adapting to climate change and variability', *Advances in Geosciences*, vol. 14, pp. 277–280.

Dahl, H, Stoltz, P & Willig, R 2004, 'Recognition, redistribution and representation in capitalist global society: an interview with Nancy Fraser', *Acta Sociologica*, vol. 47, no. 4, pp. 374–382.

Daly, H & Townsend, K 1993, *Valuing the earth: economics, ecology, ethics*, MIT Press.

Delabre, I, Boyd, E, Brockhaus, M, Carton, W, Krause, T, Newell, P, Wong, GY & Zelli, F 2020, 'Unearthing the myths of global sustainable forest governance', *Global Sustainability*, vol. 3, no. 16, pp. 1–10.

The Economist 2011, 'A man-made world: the Anthropocene', *The Economist*, Briefing section, 26 May, accessed March 2012, www.economist.com/briefing/2011/05/26/a-man-made-world.

Elliott, L 2007, 'Improving the global environment: policies, principles and institutions', *Australian Journal of International Affairs*, vol. 61, no. 1, pp. 7–14.

Evans, S & Gabbatiss, J 2019a, 'In-depth Q & A: how "Article 6" carbon markets could "make or break" the Paris Agreement', *Carbon Brief*, accessed 8 August 2020, www.carbonbrief.org/in-depth-q-and-a-how-article-6-carbon-markets-could-make-or-break-the-paris-agreement.

Evans, S & Gabbatiss, J 2019b, 'COP25: Key Outcomes agreed at the UN climate talks in Madrid', *Carbon Brief*, accessed 8 August 2020, www.carbonbrief.org/cop25-key-outcomes-agreed-at-the-un-climate-talks-in-madrid.

Filippini, A 2010, 'Women and climate change in Cochabamba', *World Rainforest Movement*, no. 154, accessed 10 June 2017, http://wrm.org.uy/bulletins/issue-154/.

Fraser, N 2005, 'Reframing global justice in a globalising world', *New Left Review*, no. 36, Nov/Dec, pp. 69–88.

Fraser, N 2008, 'Reframing justice in a globalizing world', in N Fraser, *Scales of justice: reimagining political space in a globalizing world*, Polity Press, Cambridge, UK.

Fraser, N 2020, 'Taking care of each other is essential work', Nancy Fraser interview by *VICE*, Clio Chang, 7 April 2020, accessed 1 May 2020, vice.com.

174 *Justice in the age of the Anthropocene*

Fraser, N & Olson, K 2008, *Adding insult to injury: Nancy Fraser debates her critics*, Verso, London.

Gibney, M 2000, 'Caring at a distance: (im)partiality, moral motivation and the ethics of representation – asylum and the principle of proximity', *Ethics, Place & Environment*, vol. 3, no. 3, pp. 315–317.

Ginty, A 2003, 'Citizen refugee', BSocSci (Honours) thesis, School of Social & Workplace Development, Southern Cross University, Lismore.

Global Forest Coalition 2020, '15 years of REDD+: has it been worth the money?', *Global Forest Coalition*, accessed 4 September 2020, https://globalforestcoalition.org/wp-content/uploads/2020/09/REDD-briefing.pdf.

Gray, J 2020, 'Why this crisis is a turning point in history', *New Statesman*, 1 April, accessed 15 June 2020, www.newstatesman.com/international/2020/04/why-crisis-turning-point-history.

Grayson, D & Little, B 2017, 'Conjunctural analysis and the crisis of ideas', *Soundings: A journal of politics and culture*, no. 65, Spring, pp. 59–75.

Hall, S 2011, 'The neoliberal revolution: Thatcher, Blair, Cameron – the long march of neoliberalism continues', *Soundings: A journal of politics and culture*, no. 48, Summer, pp. 9–27.

Hall, S & Massey, D 2010, 'Interpreting the crisis', *Soundings: A journal of politics and culture*, no. 44, Spring, pp. 57–71.

Harvey, D 2020, 'We need a collective response to the collective dilemma of coronavirus', *Jacobin*, 24 April, accessed 15 June 2020, https://jacobinmag.com/2020/04/david-harvey-coronavirus-pandemic-capital-economy.

Head, L 2010, 'Cultural ecology: adaptation-retrofitting a concept?', *Progress in Human Geography*, vol. 34, no. 2, pp. 234–242.

Henderson, H 1982, *Total productive system of an industrial society: layer cake with icing*, digital image, accessed 14 May 2017, http://hazelhenderson.com/wp-content/uploads/totalProductiveSystemIndustrialSociety.jpg.

Huq, S, Reid, H & Murray, L 2006, *Climate change and development links*, Gatekeeper Series 123, International Institute for Environment and Development, e-book, accessed 10 June 2017, http://lib.icimod.org/record/12247/files/4750.pdf.

Juhola, S, Glaas, E, Linner, B & Neset, T 2016, 'Redefining maladaptation', *Environmental Science & Policy*, vol. 55, January, pp. 135–140.

Ki Moon, B 2015, '"Climate change knows no national borders", UN chief says', *United Nations Sustainable Development Goals*, weblog post, n.d., accessed 21 May 2017, www.un.org/sustainabledevelopment/blog/2015/11/climate-change-knows-no-national-borders-un-chief-says/.

Kythreotis, A 2012, 'Progress in global climate change politics? reasserting national state territoriality in a "post-political" world', *Progress in Human Geography*, vol. 36, no. 4, pp. 457–474.

Lang, C 2016, 'Norway admits that "we haven't seen actual progress in reducing deforestation" in Indonesia', *REDD-Monitor*, accessed 20 June 2017, www.redd-monitor.org/2016/03/03/norway-admits-that-we-havent-seen-actual-progress-in-reducing-deforestation-in-indonesia/.

Liverman, D 2009, 'Conventions of climate change: constructions of danger and the dispossession of the atmosphere', *Journal of Historical Geography*, vol. 35, no. 2, pp. 279–296.

Lohmann, L 2008, 'Carbon trading, climate justice and the production of ignorance: ten examples', *Development*, vol. 51, no. 3, pp. 359–365.

Lohmann, L 2009, 'Regulation as corruption in the carbon offset markets', in S Bohm & S Siddhartha (eds), *Upsetting the offset: the political economy of carbon markets*, e-book, accessed 21 May 2017, http://mayflybooks.org/wp-content/uploads/2010/07/9781906948078UpsettingtheOffset.pdf.

Magnan, AK, Schipper, ELF, Burkett, M, Bharwani S, Burton, I, Erikson, S, Gemenne, F, Schaar, J & Zievogel, G 2016, 'Addressing the risk of maladaptation to climate change', *Wiley Interdisciplinary Reviews: Climate Change*, vol. 7, September/October, pp. 646–665.

Massey, D 1992, 'Politics and space/time', *New Left Review*, no. 196, Nov/Dec, pp. 65–84.

Massey, D 2011, *Spatial justice workshop – Doreen Massey*, online video, University of Westminster, accessed 3 June 2012, www.youtube.com/watch?v=l7vppCnTu88.

Mathur, V, Afionis, S, Paavola, J, Dougill, A & Stringer, L 2014, 'Experiences of host communities with carbon market projects: towards multi-level climate justice', *Climate Policy*, vol. 14, no. 1, pp. 42–62.

Maxton-Lee, B 2018, 'Narratives of sustainability: a lesson from Indonesia: global institutions are seeking to shape an understanding of sustainability that undermines its challenge to their world view', *Sounding: A journal of politics and culture*, no. 70, Winter, pp. 45–57.

Maxton-Lee, B 2020, 'Forests, carbon markets, and capitalism: how deforestation in Indonesia became a geo-political hornet's nest', Guest Post, *REDD-Monitor*, accessed 22 August 2020, https://redd-monitor.org/2020/08/21/guest-post-forests-carbon-markets-and-capitalism-how-deforestation-in-indonesia-became-a-geo-political-hornets-nest/.

McAfee, K 2014, *Green economy or buen vivir: can capitalism save itself*, Lunchtime Colloquium, Rachel Carson Centre for Environment and Society, online video, accessed 17 May 2020, www.carsoncenter.unimuenchen.de/events_conf_seminars/event_history/2014-events/2014_lc/index.html.

Mellor, M 1997, 'Women, nature and the social construction of "economic man"', *Ecological Economics*, vol. 20, no. 2, pp. 129–140.

Mellor, M 2012, *Just banking conference*, Friends of the Earth Scotland, online video, accessed 14 May 2017, www.youtube.com/watch?v=i-9vMXfTM10.

Morinière, L 2009, 'Tracing the footprint of "environmental migrants" through 50 years of literature', in A Oliver-Smith & X Shen (eds), *Linking environmental change, migration and social vulnerability*, UNU Institute for Environment and Human Security (UNU-EHS), e-book, accessed 21 May 2017, www.munichre-foundation.org/dms/MRS/Documents/Source2009_OliverSmith_ShenEnvironmentalChange_Migration.pdf.

Najam, A, Huq, S & Sokona, Y 2003, 'Climate negotiations beyond Kyoto: developing countries concerns and interests', *Climate Policy*, vol. 3, no. 3, pp. 221–231.

Okereke, C & Dooley, K 2010, 'Principles of justice in proposals and policy approaches to avoided deforestation: towards a post-Kyoto climate agreement', *Global Environmental Change*, vol. 20, no. 1, pp. 82–95.

Oktavianti, T 2020, 'Indonesia speeds up regulation on global carbon trading', *The Jakarta Post*, 7 July, accessed 16 September 2020, www.thejakartapost.com/news/2020/07/07/indonesia-speeds-up-regulation-on-global-carbon-trading.html.

Open Letter to Members of the Green Climate Fund (GCF) Board 2020, accessed 21 September 2020, https://wrm.org.uy/wp-content/uploads/2020/08/Public-Note-GCF-REDD-project-funding_Final_EN.pdf.

Osborne, T 2018, 'The de-commodification of nature: indigenous territorial claims as a challenge to carbon capitalism', *Environment and Planning E: Nature and Space*, vol. 1, no. 1–2, pp. 25–75.

176 Justice in the age of the Anthropocene

Pearce, F 2008, 'Carbon trading: dirty, sexy money', *New Scientist*, vol. 198, no. 2562, pp. 38–41.

Prins, G, Galiana, I, Green, C, Grundmann, R, Hulme, M, Korhola, A, Laird, F, Nordhaus, T, Pielke, R, Rayner, S, Sarewitz, D, Shellenberger, M, Stehr, N & Tezuka, H 2010, *The Hartwell papers: a new direction for climate policy after the crash of 2009*, London School of Economics, accessed 21 May 2017, http://eprints.lse.ac.uk/27939/1/HartwellPaper_English_version.pdf.

Röhr, U 2006, 'Gender relations in international climate change negotiations', *Berlin: LIFE eV/genanet*.

Rorty, R 1998, *Truth and progress: philosophical papers*, vol. 3, Cambridge University Press, Cambridge, UK, pp. 167–185.

Schlosberg, D 2004, 'Reconceiving environmental justice: global movements and political theories', *Environmental Politics*, vol. 13, no. 3, pp. 517–540.

Skillington, T 2012, 'Climate change and the human rights challenge: extending justice beyond the borders of the nation state', *The International Journal of Human Rights*, vol. 16, no. 8, pp. 1196–1212.

Smith, K 2008, 'Offset standard is off target', *The Corner House*, accessed 2017, www.thecornerhouse.org.uk/resource/offset-standard-target.

Smith, K 2009, 'Offset under Kyoto: a dirty deal for the south', in S Bohm & S Siddhartha (eds), *Upsetting the offset: the political economy of carbon markets*, e-book, accessed 21 May 2017, http://mayflybooks.org/wp-content/uploads/2010/07/9781906948078UpsettingtheOffset.pdf.

United Nations Framework Convention on Climate Change (UNFCCC) 1992, accessed 9 May 2017, https://unfccc.int/resource/docs/convkp/conveng.pdf.

United Nations Framework Convention on Climate Change (UNFCCC) 1998, *Kyoto Protocol to the United Nations Framework Convention on Climate Change*, United Nations, accessed 20 May 2017, https://unfccc.int/resource/docs/convkp/kpeng.pdf.

United Nations Framework Convention on Climate Change (UNFCCC) Paris Agreement 2015, accessed 16 September 2020, https://unfccc.int/process-and-meetings/the-paris-agreement/the-paris-agreement.

Van Liessum, N 2011, 'Anthropocene', *The human geography knowledge base*, wiki article, 13 September 2012, accessed 15 August 2016, http://geography.ruhosting.nl/geography/index.php?title=Environmental_Determinism.

Vigil, S 2015, 'Displacement as a consequence of climate change mitigation policies', *Forced Migration Review*, no. 49, May, pp. 43–45.

Vigil, S 2018, 'Green grabbing-induced displacement', in R McLeman & F Gemenne (eds), *Routledge Handbook of Environmental Displacement and Migration*, Routledge, London.

Wara, M 2007, 'Is the global carbon market working?', *Nature*, vol. 445, no. 7128, pp. 595–596.

Work, C, Rong, V, Song, D & Scheidel, A 2019, 'Maladaptation and development as usual? Investigating climate change mitigation and adaptation projects in Cambodia', *Climate Policy*, vol. 19, no. 51, pp. 547–562.

Wysham, D 2008, 'Carbon market fundamentalism', *Multinational Monitor*, vol. 29, no. 3, accessed 15 May 2017, www.multinationalmonitor.org/mm2008/112008/wysham.html.

Zalewski, M 2010, '"I don't even know what gender is": a discussion of the connections between gender, gender mainstreaming and feminist theory', *Review of International Studies*, vol. 36, no. 1, pp. 3–27.

Index

Page numbers in *italics* refer to figures. Page numbers followed by 'n' refer to notes.

Abbott, T 66
accumulation: by decarbonisation 32, 66, 135, 160; by dispossession 32, 68
adaptation 5–8; actions to climate change 36; community-based 42; conceptual evolution of 8; plus development 7–8, 41–42, 159; and development policy nexus 8; equal footing 34, 36; international policy 34–39; overview of 28–34; policy 18, 38, 39, 159, 166–169; precautious 16, 29, 31, 38–39, 40, 168; stand-alone 40–41; sustainable 31, 36, 37, 39, 160; taboo on 40; types of 34, 39–42
adaptive capacity 35
additionality: concept of 55–56, 162, 169; and counterfactuals 79, 81; legitimacy 68; tool 58; types 56, 162; verifiability of 58
Ad Hoc Working Group on Long-term Cooperative Action (AWG-LCA) 92
agency 157
agrarian control, human security and 136–140
Aliansi Masyarakat Adat Nusantara (AMAN) 104, 112, 168, 169
AMAN Central Kalimantan 104
Annex B Parties, under Kyoto Protocol 55–56
Anthropocene 172n1
Arora-Jonsson, S 121, 128
Article 2 of Climate Convention 7, 8, 33, 35, 54, 157–159
Article 3.3 of UNFCCC 31, 54
Article 6.4 of Paris Agreement 21, 47, 50, 60, 65, 145, 151, 163
Association of Palm Oil Plantations in Indonesia 165

AusAID website 107
Australia Indonesia Partnership (AIP) 107
Ayers, J 7, 34, 40, 41

Baier, A 147
Baldwin, A 17
Bank Information Centre 104
Bantaya 104
Barnett, C 147, 158
Barnett, J 1, 2, 4, 29, 30, 31, 33, 53, 55, 70, 160, 161
Beijing Women's Conference (1995) 123, 124, 128, 150
Bennholdt-Thomsen, V 84
Berger, J 36
Bessis, S 8–9, 87–90, 95, 140n1
Bettini, G 17
Bond, P 33
Borneo Orang-utan Survival (BOS Mawas) 110–112
Bosquet, B 82
Brazil 60
Bretton Woods system 88
Brooks, N 33, 39, 41
Brown, K 37, 39, 160
Bumpus, A 32, 48, 66, 80
Burton, I 34, 40
Business Case for Mainstreaming Gender in REDD+, The 92

Cambodia: CCMA projects in 4, 145, 159–160; irrigation and reforestation in 103
Cames, M 51, 65, 162, 163
Cancun Adaptation Framework 47
Cancun Agreements 92
capital accumulation 68, 84, 170

178   *Index*

capitalism 135; carbon 2, 9, 68, 91, 102, 120, 151; global 85
carbon: capitalism 2, 9, 68, 91, 102, 120, 151; commodification 91, 151, 167; markets 131–132; offset 48; taxes 59, 66, 67, 160; trading, regulation on 152
Carbon Market Watch 59
CARE International 110, 111
Center for International Forestry Research's (CIFOR) review 120, 134, 150
Central Kalimantan, Indonesia: outcomes and analysis 107–116; UN-REDD+ study 103, 106, 163
Central Kalimantan Peatlands Project (CKPP) 110, 111
Central Sulawesi, Indonesia: outcomes and analysis 116–117; UN-REDD+ study 103, 106, 163, 164
certified emission reductions (CERs) 3, 49, 60; credits 54, 56, 57, 162–163; volume 62, 64
Chile 60
China 60, 163
Ciba 138
classical liberal feminism 16–17, 21, 151
Clean Development Mechanism (CDM): Article 12.2/12.5 of Kyoto Protocol 48, 55–56; available choices to future generations (path dependency) 69–70; CERs volume, in Asia/Latin America 62, 64; comprehensive analysis of 3; disproportionately burdening the most vulnerable 60; emission reductions under 162; environmental integrity in 159; high opportunity costs of 65–68, 160; increasing emissions of greenhouse gases 55–60, 57; introduction 48–53; Kyoto Protocol 2, 9, 18, 19, 20, 32, 37, 48, 55–56, 134, 162; and maladaptation 52, 53–55, 160; as model for REDD in climate policy strategy 78, 81; projects, in Asia/Latin America 61, 63; reduces incentives to adapt 68–69; sustainable development 37, 48–49, 51, 65, 67
climate: justice 133, 138, 139, 152, 155; migration 3; proofing 159; stability 51, 158
climate change mitigation and adaptation (CCMA) initiatives 4, 103, 145
climate change solutions: adaptation and mitigation 2–3, 5–8; analysing technical nature of 18; critiques of 18; leading to maladaptation 161–166; perversity of 9; politics of 19

Climate Convention (1992) 7, 8, 33, 35, 54, 157–159
Climate Investment Funds (CIFs) 90
cognitive lock-in 147
'common but differentiated responsibilities' (CBDR) principle 5, 133, 147, 157–158
community-based adaptation (CBA) 42
community monitoring 135, 136, 161, 168
conceptual research 16; data collection and analysis 20–23; research design 17–20; using critical theoretical methodology 17–23
Conference of Parties (COP): in Bali (2007) 34–35, 79, 87, 89, 92, 96n2; in Cancun (2010) 47, 92; in Doha (2012) 54; in Glasgow (2021) 50, 51; in Kyoto (1997) 140n1; in Madrid (2019) 50, 51; in Poland (2018) 50; UNFCCC 5–8, 19, 34, 47, 69, 87, 133, 136, 157, 159, 166
conjunctural crisis 84, 169
Convention on the Elimination of All Forms of Discrimination Against Women 88
Coopa 138
Copenhagen Accord (2009) 77, 147
cosmopolitanism 146
COVID-19 pandemic 84
critical theoretical research 16, 17
critical theory versus positivism 16
Crutzen, P 172n1
cultural injustice (misrecognition) 148, 150, 168

Daly, H 82, 96n1, 160, 169
Dayak tribes 127
debtWatch 104
decarbonisation 47, 59; accumulation by 32, 66, 135, 160; dispossession by 80
deforestation and forest degradation, reducing emissions from 20, 79, 84, 107, 115, 126; *see also* United Nations Programme on Reducing Emissions from Deforestation and Forest Degradation (UN-REDD+) programmes
de-gendered economics 79–91
Delabre, I 147
democratic representation 149
development forced displacement and resettlement (DFDR) 163
displacement 4, 47, 105, 106, 145, 163
dispossession: by accumulation 32, 68; by decarbonisation 80
Dodman, D 7, 34, 40, 41
Downer, A 109

Index 179

Earth Summit 7
ecofeminism 16, 20, 21, 82–87, 151
ecological limits 6, 35
economic: dualism 86, 87, 90, 91, 95, 151–153, 161; injustice 148; redistribution (maldistribution) 144, 149, 151, 168
efficiency/efficacy, increasing 93, 94
Elliott, L 31, 36, 147
emancipatory thrust 18
emission reduction instrumentation 132
Emissions Trading and Joint Implementation 48, 69
Emissions Trading Scheme (ETS) 67
emission-trading system 152
environment/environmental: additionality 56, 162; as anthropogenic climate change 157–158; co-benefits 79; determinism 30, 32, 157; geography 29–30; injustices 152, 158; justice 133, 138, 139, 152, 155–156; migration 3–5, 145; reintegration 146 150, 155–161; see also climate
Environmental Change and Forced Migration Scenarios (EACH-FOR) 10n1
Environmental Emergency Migrants/ Displacees 10n1
environmental integrity: in CDM 159; as manifestation of high environmental costs 65; and social justice 36, 37, 39–40, 160
Environmentally Forced Migrants 10n1
environmentally induced migration displacement 4, 10n1, 103, 161–166
Environmentally Motivated Migrants 10n1
Eriksen, S 28, 37
essentialism 83, 85
ETS (Emissions Trading Scheme) 67
Ex-Mega Rice Project (EMRP) 108–110, 152
Extinction Rebellion (XR) 171

feminism 8, 77, 88; feminist qualitative research, in Indonesia 23–25, 103–105; intra-disciplinary differences between 16; see also classical liberal feminism; ecofeminism; women
financial additionality 56, 162
Food and Agriculture Organization (FAO) 20, 81, 92
Forest Carbon Partnership Facility (FCPF) 81–82, 89, 92, 96n2–3, 104

forest degradation see deforestation and forest degradation
Forest Plantation and the Palm Oil Association 165–166, 168
Forestry Act 113
foundational thinking 147
Fourth World Conference of Women, in Beijing (1995) 123, 124, 128, 150
'fragile states' 81
Fraser's three-dimensional theory of justice 9–10, 144, 145, 147–150, 154–156, 158, 166–169
free-market environmentalism 55
free-market neoliberalism 167
Free Prior and Informed Consent (FPIC) 106, 108, 110, 116, 134, 150, 164
Friends of the Earth Australia 114

gender 8–9, 22; equality 8, 22, 88, 89, 93–95, 123–124, 126, 150; equity 89; gap 18, 79; justice 133, 139, 152; neutrality 140, 152, 159; in operationalising UN-REDD+ policies 92–95; see also feminism; women
Gender Action Plan (2006) 89
gendered silent offset economy 2, 17, 102, 120, 151–153
gender mainstreaming 17, 21–22; carbon markets and 132; meaningful participation and 153; in REDD+ 24, 93–94, 120, 122–131
geography 39
Georgescu-Roegen, N 90
Gillard, J 66
global capitalism 85;
global carbon market 48, 60, 82, 105, 166
global carbon mitigation programme 20
global gender equality 8, 88
Global Deal, The (Stern) 82
Global Environmental Facility (GEF) 41
Global Financial Crisis (GFC) of 2008 84
Global Forest Coalition 120, 134, 165
Gore, Al 6
Grambsch, A 30
Gray, J 37, 38, 170
Green Climate Fund (GCF) 152–153
green-grabbing displacement 4, 164, 166
greenhouse gas (GHG) emissions: fictitious commodity in 67–68; increase in 55–60; indulgence system, modern day 53; limitation of crediting mechanisms 65; reductions 35, 48–49, 55, 57, 67, 93, 94, 129, 152
Green Revolution 138, 152

180   *Index*

Hall, S 84, 170
Hansen, G 138
Hansen, J 66, 69
Harvey, D 32, 68, 84, 170
Head, L 32, 157
Heiman, M 53
Henderson, H 84, 91
high opportunity costs, of CDM 65–68, 160
Howard, J 107
HuMa 104, 105, 116, 122
human agency 20, 157, 158
human mobility 4
human security and agrarian control 136–140
Huq, S 36, 40, 159, 167
hydroelectric dams 163

Iceberg Model 85, 86, 91, 151
ideational crisis 146
impossibility theorems 58
impoverishment 138
incentives theory, w. r. t. REDD+ mechanism 81
Independent Evaluation Group (IEG) 92, 96n2
India 60
Indigenous Environmental Network (IEN) 82
Indigenous Organisations of the Amazon River Basin (COICA) 161
Indigenous Peoples Alliance of the Archipelago: Aliansi Masyarakat Adat Nusantara (AMAN) 104, 112, 168, 169; AMAN Central Kalimantan 104; Perempuan AMAN (AMAN Women) 104
indigenous women, UN-REDD+ impacts on 23–25, 77–95; *see also* Indonesian study; women
Indonesia-Australia Forest Carbon Partnership (IAFCP) 107
Indonesian Association for Forest Corporation 116
Indonesian study: Central Kalimantan, KFCP (outcomes and analysis) 107–116; Central Sulawesi (outcomes and analysis) 116–117; discussion of findings 122–140; findings of 102–118; gendered impacts of UN-REDD+ 122–131; major findings (for recognition of adat tenure) 153–155; outcomes of 105–117; rationale of 103; research methodology 103–105; summary of outcomes of 105–107
innovative market mechanisms 32, 145

Intergovernmental Panel on Climate Change (IPCC) 5, 7, 8, 19, 29, 35, 36, 53–54, 79, 150, 159
international adaptation policy 34–38
International Rivers 57, 163

Jackson, C 83
Jakarta 103–104
Joint Implementation and Emissions Trading 48, 69
Juhola, S 2, 5, 29, 53
justice 9–10, 19, 132–133; in age of Anthropocene 144–172; background thoughts on 146–161; climate/gender 133, 138, 139, 152, 155; environmental reintegration and injustice of maladaptation 150, 155–161; Fraserian three-dimensional framework 147–150; misframing 149; parity of participation 148, 150; problematising/thematising 148, 158

Kalimantan Forests and Climate Partnership (KFCP) project 104, 106, 107–116, 121, 126, 130, 150, 163
Karsenty, A 81
'Keynesian-Westphalian frame' phrase 149, 172n2
Kyoto Protocol: Adaptation Fund 32, 36, 54, 69; *see also* Clean Development Mechanism (CDM)

land rights 136, 156
'Layer Cake with Icing' 85, 85, 91, 151
Leach, M 83, 88
Least Developed Countries Fund (LDCF) 41
Lembaga Dayak Panarung 104
Libu Perempuan 104
limits of tolerance 6, 35
Liverman, D 30, 32, 48, 58, 66, 67, 80
Lohmann, L 38
Lovelock, J 37, 135

Mace, M 7
MacGregor, S 79
Magnan, AK 1, 16, 29, 158–159
maladaptation: in action 103; adaptation policy and 28–31, 159; addressing 5; CDM and 52, 53–55; defined 1–2, 5, 36; framing of 159; identification 10; injustice of 146, 150, 155–161; justice and 10; key terms/concepts links to 2–10; manufacturing 47–70; typology 33, 53, 160, 161

Index 181

Maladaptation Fund 69
maldistribution, injustice of 148, 149, 151, 152, 168
Malthus, T 96n1
market environmentalism 68, 69, 161; *see also* neoliberal market environmentalism
Massey, D 84, 85, 169, 170
masyarakat adat 104, 121, 122
Maxton-Lee, B 135, 137
McAfee, K 78, 94
McCarthy, J 51
meaningful participation 124, 126, 132, 150, 153, 156
Mellor, M 80, 84, 86–87, 90, 91, 95, 96, 151, 152, 161
Mies, M 84, 85
Ministry of Forestry, Indonesia 165, 166
misrecognition, injustice of 148, 150, 168
misrepresentation (political injustice) 148, 149, 168
mitigation 5; bias toward 6, 8, 19; climate change policy 47; and Kyoto CDM 47–70; policy 18, 33, 37, 52, 70, 166–169
monetary value 67

National Adaptation Programmes of Action (NAPAs) 41
National REDD+ Strategy 123, 126
natural capital 96n1
neoliberalism 18, 51–53, 80, 84, 85, 87
neoliberal market environmentalism 2, 16, 21, 36–38, 52, 55, 134, 147, 166; *see also* market environmentalism
Norway 168–169

offsets 79; carbon 48; silent 77, 78, 86, 91–95, 150–151
O'Neill, S 1, 2, 4, 29, 30, 31, 33, 53, 55, 70, 160, 161
Ongolo, S 81
opportunity cost 65–68, 81

Palangkaraya, in Central Kalimantan 103–104
Palm Oil Association 165–166, 168
Palu, in Central Sulawesi 103–104
Paris Agreement (Article 6.4) 21, 47, 50, 60, 65, 145, 151, 163
Parreno, J 49
Parties to the Paris Agreement (CMA), meeting of 50–51
Paulsson, E 50, 58

Payment for Environmental Services (PES) 67, 78, 134
Pearce, F 57
Perempuan AMAN (AMAN Women) 104
Pielke, R 6, 47
'Pivot to Extinction' diagram 170–171, *171*
political injustice (misrepresentation) 148, 149, 168
political representation 149, 153, 154
political subsurface 17
polluter pays principle (PPP) 133–136, 141n2, 157
positivism 20
post-Sukarno 'Green Revolution' campaign 138
precautionary principle (PP) 31, 33, 35, 36, 38–39, 54, 78, 159
precautious adaptation 16, 29, 31, 38–39, 40, 168
primitive accumulation 32, 68, 70n2
Prins, G 5, 66
'provisioning' 86, 90
Prudham, S 51

Rayner, S 5
RDD&D 67
Reagan, R 84
recognition 149, 153–155, 156, 157, 167
redistribution 149, 156, 167
reintegration, as fourth dimension of justice 150, 155–161, 168
representation 148, 149, 156, 167
Revenge of Gaia (Lovelock) 135
Ricardo, D 96n1
Rorty, R 147

Sandel, M 80
Scheraga, J 30
Schipper, ELF 6, 7, 35, 38, 40
Schlosberg, D 147, 152, 155–156
self-reflection 24
slippery commodity, of CDM 67, 69
Smith, A 96n1
social inequality 148, 150
socialisation and FPIC 116
social justice 10, 36, 37, 39–40, 160
social order 68, 158
social reality 17, 20
social reproduction 86, 91, 95, 102, 120, 151
social science 30
Solidaritas Perempuan 104
Solomon, B 53
stand-alone adaptation 40–41
Stern, N 32, 43n1, 82

182 Index

Stern Report on Climate Change (2006) 32
Stoemer, E 172n1
Subsidiary Body on Scientific and
  Technical Advice (SBSTA) 35
Suharto 138
sustainability 93, 94, 146
sustainable adaptation 31, 36, 37, 39, 160
sustainable development, in CDM offset
  projects 37, 48–49, 51, 65, 67
sustainable growth 37, 39, 160
Sutter, C 49

Thatcher, M 84
three-dimensional justice framework,
  Fraser's 9–10, 144, 145, 147–150,
  154–156, 158, 166–169
Tickell, C 3
time, irreversibility of 90
tolerable limits 6, 35
total productive system, of industrial
  society 85
Townsend, K 160, 169
transaction costs 67, 94

United Nations: Conference for Environment and Development (1992), in Rio
  de Janeiro 7; Declaration of the Rights
  of Indigenous Peoples 110; Development Programme (UNDP) 20, 81, 89,
  92; Environment Programme (UNEP)
  20, 81, 92; Food and Agriculture
  Organization (FAO) 20, 81, 92
United Nations Framework Convention
  on Climate Change (UNFCCC): adaptation and mitigation policies under 3,
  4, 6, 8, 16, 19, 32, 33, 38, 40, 145, 161,
  166–169; Article 3.3 of 31, 54; CBDR
  principle in 157; Climate Convention
  (1992) 7, 8, 33, 35, 54, 157–159; concept of additionality 55–56; Conference
  of the Parties (COP) 5–8, 19, 34, 47,
  69, 87, 133, 136, 157, 159, 166; interpretation of adaptation 28–29, 34, 37,
  41; IPCC and 29, 30; Kyoto CDM 9,
  32, 47, 77, 79, 147; mitigation bias 34;
  SBSTA 35
United Nations Programme on Reducing
  Emissions from Deforestation and
  Forest Degradation (UN-REDD+) programmes 2, 9, 20–22, 77–79, 96n2, 145,

163–166, 168–169; critics of 79; gender
  analysis, in policies 92–95; gendered
  impacts on women 122–131; impacts
  on indigenous women/perempuan adat
  23–25, 77–91, 121; operationalising
  REDD+ 91–95; Readiness Fund 92;
  and World Bank 81, 87–91
Universal Declaration of Human Rights 88
University of Palangka Raya (UPR)
  110, 111
unsustainable development 36, 37, 39,
  159, 167

Vandana Shiva 85
Victor, D 57, 162
Vigil, S 4, 145, 164, 166
vulgar economism 167
vulnerability 2, 30, 53

WAHLI (Friends of the Earth, Indonesia)
  114
Wetlands 111
Whiteside, K 38
women: capacity-building activities 126;
  in climate change discourse 8–9, 16–17,
  120; environmental policy and 121;
  environmental resources and 89; environment and development (WED) 83,
  88; exclusion/inclusion in forestry
  sector 93, 121; gendered impacts of
  REDD+ on 122–131; indigenous,
  UN-REDD+ impacts on 23–25, 77–95;
  silence/invisibility in UN-REDD+ 77,
  79–95, 151–152; as silent offsets in
  developing countries/global South 77,
  78, 86, 91–95, 150–151; UN-REDD+
  impacts on 23–25, 77–95; see also
  feminism; gender; Indonesian study
Women Organizing for Change in
  Agriculture and Natural Resource
  Management (WOCAN) 93, 121
Woods Hole Research Centre 161
Work, C 2, 4, 103, 145, 159
World Bank (WB) 78, 81–82, 87–90, 92,
  96n2, 104, 152
World Rainforest Movement 84
WWF 111

Yayasan Petak Danum Kalimantan
  Tengah (YPD) 111–114

Printed in the United States
by Baker & Taylor Publisher Services